BIOCHEMICAL SOCIETY SYMPOSIA

No. 56

G-PROTEINS AND SIGNAL TRANSDUCTION

BIOCHEMICAL SOCIETY SYMPOSIUM No. 56
held at St Bartholomew's Hospital Medical College,
London, December 1989

G-Proteins and Signal Transduction

ORGANIZED AND EDITED BY

G. MILLIGAN, M. J. O. WAKELAM
AND J. KAY

1990
LONDON: THE BIOCHEMICAL SOCIETY

Published by Portland Press, 59 Portland Place, London W1N 3AJ
on behalf of the Biochemical Society

ISBN: 1 85578 001 1
ISSN: 0667 8694

Typeset by Santype International Limited, Salisbury and
Printed in Great Britain by Dotesios Printers Ltd

List of Contributors

T. K. Attwood (*Department of Biochemistry, University of Leeds, Leeds LS2 9JT, U.K.*)

P. Benoît (*Institut Pasteur, Laboratoire de Neurobiologie Moleculaire, 25 rue du Dr Roux, 75724 Paris Cedex 15, France*)

A. Bessis (*Institut Pasteur, Laboratoire de Neurobiologie Moleculaire, 25 rue du Dr Roux, 75724 Paris Cedex 15, France*)

F. M. Black (*Molecular Pharmacology Group, Department of Biochemistry, University of Glasgow, Glasgow G12 8QQ, Scotland, U.K.*)

A. A. Bominaar (*Department of Biochemistry, University of Groningen, Nijenborgh 16, 9747 AG Groningen, The Netherlands*)

M. Bushfield (*Molecular Pharmacology Group, Institute of Biochemistry, University of Glasgow, Glasgow G12 8QQ, Scotland, U.K.*)

J. Cartaud (*Laboratoire de Microscopie Electronique, Institut Jacques Monod, Université Paris 7, 2 place Jussieu, Tour 43, 75251 Paris Cedex 05, France*)

J. P. Changeux (*Institut Pasteur, Laboratoire de Neurobiologie Moleculaire, 25 rue du Dr Roux, 75724 Paris Cedex 15, France*)

Y. Churcher (*Department of Physiology, University College London, University Street, London WC1E 6JJ, U.K.*)

S. A. Davies (*Molecular Pharmacology Group, Department of Biochemistry, University of Glasgow, Glasgow G12 8QQ, Scotland, U.K.*)

A. Devillers-Thiéry (*Institut Pasteur, Laboratoire de Neurobiologie Moleculaire, 25 rue du Dr Roux, 75724 Paris Cedex 15, France*)

I. Diamond (*Ernest Gallo Clinic and Research Center, San Francisco General Hospital, San Francisco, CA 94110, U.S.A. and Departments of Neurology, Pediatrics and Pharmacology, University of California, San Francisco, CA 94143, U.S.A.*)

A. C. Dolphin (*Department of Pharmacology, St George's Hospital Medical School, Cranmer Terrace, London SW17 0RE, U.K.*)

E. E. Eliopoulos (*Astbury Department of Biophysics, University of Leeds, Leeds LS2 9JT, U.K.*)

J. B. C. Findlay (*Department of Biochemistry, University of Leeds, Leeds LS2 9JT, U.K.*)

B. Fontaine (*Institut Pasteur, Laboratoire de Neurobiologie Moleculaire, 25 rue du Dr Roux, 75724 Paris Cedex 15, France*)

LIST OF CONTRIBUTORS

J. L. Galzi (*Institut Pasteur, Laboratoire de Neurobiologie Moleculaire, 25 rue du Dr Roux, 75724 Paris Cedex 15, France*)

P. K. Goldsmith (*Molecular Pathophysiology Branch, National Institute of Diabetes, Digestive and Kidney Diseases, National Institutes of Health, Bethesda, MD 20892, U.S.A.*)

B. D. Gomperts (*Department of Physiology, University College London, University Street, London WC1E 6JJ, U.K.*)

A. S. Gordon (*Ernest Gallo Clinic and Research Center, San Francisco General Hospital, San Francisco, CA 94110, U.S.A. and Departments of Neurology and Pharmacology, University of California, San Francisco, CA 94143, U.S.A.*)

S. L. Griffiths (*Molecular Pharmacology Group, Institute of Biochemistry, University of Glasgow, Glasgow G12 8QQ, Scotland, U.K.*)

J. R. Hadcock (*Department of Pharmacology, Diabetes and Metabolic Diseases Research Program, State University of New York at Stony Brook, Stony Brook, NY 11794, U.S.A.*)

H. E. Hamm (*Department of Physiology and Biophysics, University of Illinois College of Medicine at Chicago, Box 4998, Chicago, IL 60680, U.S.A.*)

M. D. Houslay (*Molecular Pharmacology Group, Institute of Biochemistry, University of Glasgow, Glasgow G12 8QQ, Scotland, U.K.*)

J. B. Hurley (*Howard Hughes Medical Institute, SL-15, University of Washington, Seattle, WA 98195, U.S.A.*)

E. Huston (*Department of Pharmacology, St George's Hospital Medical School, Cranmer Terrace, London SW17 0RE, U.K.*)

T. L. Z. Jones (*Molecular Pathophysiology Branch, National Institute of Diabetes, Digestive and Kidney Diseases, National Institutes of Health, Bethesda, MD 20892, U.S.A.*)

F. Kesbeke (*Cell Biology and Genetics Unit, Zoological Laboratory, Kaiserstraat 63, 2311 GP Leiden, The Netherlands*)

A. Klarsfeld (*Institut Pasteur, Laboratoire de Neurobiologie Moleculaire, 25 rue du Dr Roux, 75724 Paris Cedex 15, France*)

A. Koffer (*Department of Physiology, University College London, University Street, London WC1E 6JJ, U.K.*)

I. M. Kramer (*Department of Physiology, University College London, University Street, London WC1E 6JJ, U.K.*)

R. Laufer (*Institut Pasteur, Laboratoire de Neurobiologie Moleculaire, 25 rue du Dr Roux, 75724 Paris Cedex 15, France*)

B. Lavan (*Molecular Pharmacology Group, Institute of Biochemistry, University of Glasgow, Glasgow G12 8QQ, Scotland, U.K.*)

LIST OF CONTRIBUTORS

T. Lillie (*Department of Physiology, University College London, University Street, London WC1E 6JJ, U.K.*)

F. R. McKenzie (*Molecular Pharmacology Group, Department of Pharmacology, University of Glasgow, Glasgow G12 8QQ, Scotland, U.K.*)

C. C. Malbon (*Department of Pharmacology, Diabetes and Metabolic Diseases Research Program, State University of New York at Stony Brook, Stony Brook, NY 11794, U.S.A.*)

J. Malinski (*Department of Physiology and Biophysics, University of Illinois College of Medicine at Chicago, Box 4998, Chicago, IL 60680, U.S.A.*)

M. Mazzoni (*Department of Physiology and Biophysics, University of Illinois College of Medicine at Chicago, Box 4998, Chicago, IL 60680, U.S.A.*)

G. Milligan (*Molecular Pharmacology Group, Department of Biochemistry, University of Glasgow, Glasgow G12 8QQ, Scotland, U.K.*)

D. Mochly-Rosen (*Ernest Gallo Clinic and Research Center, San Francisco General Hospital, San Francisco, CA 94110, U.S.A. and Departments of Neurology and Pharmacology, University of California, San Francisco, CA 94143, U.S.A.*)

I. Mullaney (*Molecular Pharmacology Group, Department of Biochemistry, University of Glasgow, Glasgow G12 8QQ, Scotland, U.K.*)

C. Mulle (*Institut Pasteur, Laboratoire de Neurobiologie Moleculaire, 25 rue du Dr Roux, 75724 Paris Cedex 15, France*)

L. Nagy (*Ernest Gallo Clinic and Research Center, San Francisco General Hospital, San Francisco, CA 94110, U.S.A. and Department of Neurology, University of California, San Francisco, CA 94143, U.S.A.*)

H. O. Nghiêm (*Institut Pasteur, Laboratoire de Neurobiologie Moleculaire, 25 rue du Dr Roux, 75724 Paris Cedex 15, France*)

M. Osterlünd (*Institut Pasteur, Laboratoire de Neurobiologie Moleculaire, 25 rue du Dr Roux, 75724 Paris Cedex 15, France*)

J. Piette (*Institut Pasteur, Laboratoire de Neurobiologie Moleculaire, 25 rue du Dr Roux, 75724 Paris Cedex 15, France*)

P. J. Rapiejko (*Department of Pharmacology, Diabetes and Metabolic Diseases Research Program, State University of New York at Stony Brook, Stony Brook, NY 11794, U.S.A.*)

H. Rarick (*Department of Physiology and Biophysics, University of Illinois College of Medicine at Chicago, Box 4998, Chicago, IL 60680, U.S.A.*)

F. Revah (*Institut Pasteur, Laboratoire de Neurobiologie Moleculaire, 25 rue du Dr Roux, 75724 Paris Cedex 15, France*)

M. Ros (*Department of Pharmacology, Diabetes and Metabolic Diseases Research Program, State University of New York at Stony Brook, Stony Brook, NY 11794, U.S.A.*)

J. Schlessinger (*Rorer Biotechnology Inc., 680 Allendale Road, King of Prussia, PA 19406, U.S.A.*)

R. H. Scott (*Department of Pharmacology, St George's Hospital Medical School, Cranmer Terrace, London SW17 0RE, U.K.*)

Y. Shakur (*Molecular Pharmacology Group, Institute of Biochemistry, University of Glasgow, Glasgow G12 8QQ, Scotland, U.K.*)

W. F. Simonds (*Molecular Pathophysiology Branch, National Institute of Diabetes, Digestive and Kidney Diseases, National Institutes of Health, Bethesda, MD 20892, U.S.A.*)

E. Snaar-Jagalska (*Cell Biology and Genetics Unit, Zoological Laboratory, Kaiserstraat 63, 2311 GP Leiden, The Netherlands*)

A. M. Spiegel (*Molecular Pathophysiology Branch, National Institute of Diabetes, Digestive and Kidney Diseases, National Institutes of Health, Bethesda, MD 20892, U.S.A.*)

D. Strassheim (*Molecular Pharmacology Group, Institute of Biochemistry, University of Glasgow, Glasgow G12 8QQ, Scotland, U.K.*)

K.-H. Suh (*Department of Physiology and Biophysics, University of Illinois College of Medicine at Chicago, Box 4998, Chicago, IL 60680, U.S.A.*)

E. Tang (*Molecular Pharmacology Group, Institute of Biochemistry, University of Glasgow, Glasgow G12 8QQ, Scotland, U.K.*)

P. E. R. Tatham (*Department of Physiology, University College London, University Street, London WC1E 6JJ, U.K.*)

C. G. Unson (*Department of Biochemistry, Rockefeller University, New York, NY 10021, U.S.A.*)

L. Vallar (*Department of Pharmacology, C.N.R. Centre of Cytopharmacology, Scientific Institute San Raffaele, University of Milan, Milan 20132, Italy*)

J. Van der Kaay (*Department of Biochemistry, University of Groningen, Nijenborgh 16, 9747 AG Groningen, The Netherlands*)

P. J. M. Van Haastert (*Department of Biochemistry, University of Groningen, Nijenborgh 16, 9747 AG Groningen, The Netherlands*)

M. J. O. Wakelam (*Molecular Pharmacology Group, Department of Biochemistry, University of Glasgow, Glasgow G12 8QQ, Scotland, U.K.*)

H.-Y. Wang (*Department of Pharmacology, Diabetes and Metabolic Diseases Research Program, State University of New York at Stony Brook, Stony Brook, NY 11794, U.S.A.*)

D. C. Watkins (*Department of Pharmacology, Diabetes and Metabolic Diseases Research Program, State University of New York at Stony Brook, Stony Brook, NY 11794, U.S.A.*)

Preface

Guanine-nucleotide-binding proteins (G-proteins) are a ubiquitous family of proteins involved in the transmission of hormone, neurotransmitter and growth factor encoded information across the plasma membrane (and potentially other membrane-limited compartments) of cells. The remarkable conservation of sequence of the individual G-proteins throughout evolution, from insects and slime moulds to mammals, has indicated the potential for the use of the sophisticated genetics available in such systems to define the specific roles of each G-protein. Further, the generation of probes for the various G-proteins, and their corresponding mRNAs, has led to demonstrations of mutations or alterations in either the function or expression of a number of G-proteins in a range of clinical disorders, as well as an understanding of the details of how G-proteins interact with both receptor and effector systems. In this book, acknowledged experts in the field provide a timely discussion of the structure, diversity, molecular biology and regulation of expression and function of the G-proteins and other components of signal transduction cascades.

G. MILLIGAN
M. J. O. WAKELAM
J. KAY

Glasgow and Cardiff

Contents

CONTENTS

Abbreviations

AChR	Acetylcholine receptor
ADA	Adenosine deaminase
β-ARK	β-Adrenergic-receptor-specific kinase
CAT	Chloramphenicol acetyl transferase
CGRP	Calcitonin gene-related peptide
DADLE	[D-Ala^2, Leu^5]Enkephalin
DAG	*sn*-1,2-Diacylglycerol
DALAMID	[D-Ala^2]Leucine enkephalinamide
DRG	Dorsal root ganglion
EGF	Epidermal growth factor
ELISA	Enzyme-linked immunosorbent assay
G_t	Transducin
$GABA_B$	γ-Aminobutyric $acid_B$
GAP	GTPase-activating protein
GH	Growth hormone
GHRH	Growth hormone-releasing hormone
Gpp[CH_2]p	Guanosine 5′-[αβ-methylene]triphosphate
Gpp[NH]p	Guanosine 5′-[βγ-imido]triphosphate
GRE	Glucocorticoid response element
GroPCho	Glycerophosphocholine
GTP[S]	Guanosine 5′-[γ-thio]triphosphate
IBMX	3-Isobutyl-1-methylxanthine
Ins*P*	*myo*-Inositol; *P*, phosphate with locants designated where appropriate
mAb	Monoclonal antibody
MMTV-LTR	Mouse mammary tumour virus–long terminal repeat
OAG	Oleoylacetylglycerol
PDE	Phosphodiesterase
PDGF	Platelet-derived growth factor
PIA	N^6-Phenylisopropyladenosine
PKC	Protein kinase C
PG	Prostaglandin
PMA	See TPA
PtdCho	Phosphatidylcholine
PtdIns	Phosphatidylinositol
PTK	Protein tyrosine kinase
R*	Light-activated rhodopsin
SAH	*S*-Adenosyl homocysteine
TPA	12-*O*-Tetradecanoylphorbol 13-acetate
TTX	Tetrodotoxin

Biochem. Soc. Symp. **56**, 1–8
Printed in Great Britain

The Structure of G-Protein-Linked Receptors

JOHN B. C. FINDLAY, ELIAS E. ELIOPOULOS† and TERESA K. ATTWOOD

The Department of Biochemistry and †The Astbury Department of Biophysics, University of Leeds, Leeds LS2 9JT, U.K.

Introduction

It is becoming increasingly likely that membrane-bound receptors which interact with guanine-nucleotide-binding proteins represent the most widespread and prolific information-transfer system in eukaryotic biology. Growth in the number of receptors belonging to this family is entering an exponential phase, but although we now have a great deal of information on the diversity of their primary structures, this is not matched by any definitive knowledge on those sites on the proteins that confer their distinctive specificities. To arrive at some appreciation of the three-dimensional structure of an integral membrane protein, we had already embarked on an extensive protein chemistry study of the 'visual pigment' rhodopsin, soon to become one of the key members of the group. This paper will briefly describe the structural model that emerged from this study and its extrapolation to other receptors of this type. The information gained from the comparative analysis of the various subgroups in the family, in the context of the structural model, has shed some light on possible critical structural features of the protein and on important interactive sites in this structure.

Structural Model for Rhodopsin

The identification, by protein sequencing, of a large number of sites on rhodopsin subject to chemical or biochemical modification or to cleavage by proteolytic enzymes, established that the protein traversed the lipid bilayer seven times (reviewed in [1,2]). The resulting disposition of the polypeptide chain with respect to the membrane, placed the attachment point for the chromophore retinal (Lys-296 on the seventh transmembrane segment) deep within the confines of the bilayer, an important observation to which we will return. The application of structure prediction algorithms [3] and hydropathy plots suggested that the seventh transmembrane segment was neither a completely regular helix nor convincingly hydrophobic, perhaps indicative of an important functional role.

At that time the helix-packing arrangement of another protein, bacteriorhodopsin, possessing seven transmembrane segments, had been elucidated [4]. It may be coincidental that both proteins utilize light via the same

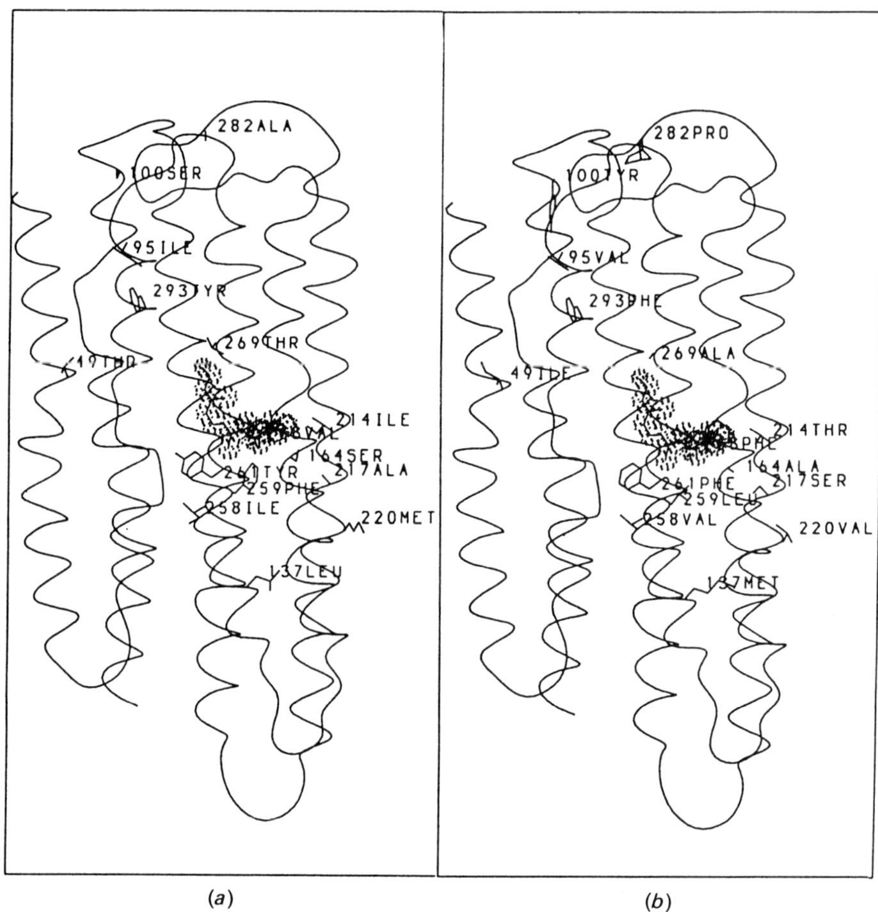

Fig. 1. *Representations of the red (a) and green (b) human colour pigments modelled on the structure for rhodopsin*

Side-chains shown are those that vary between the two colour pigments. Also shown is a putative location for the chromophore. The extracellular surface is at the top of the diagram.

chromophore, attached in the same way at a very similar point in the polypeptide chain, for their activities and sequences show no homologous relationships. More importantly, biophysical data tended to suggest that the cross-sectional areas of the two molecules in the bilayer were almost identical. To a first approximation, therefore, it seemed reasonable to assume that the gross helix-packing arrangement of the two proteins was similar. The electron density profile for bacteriorhodopsin (kindly provided by R. Henderson), thus gave not only the relative positions of the seven helices, but also their angular interrelationships. Thereafter, the model was constructed from the sequence of rhodopsin, from the intramembranous availability of side-chains to modification [5], from structure prediction data and paying particular attention to helix connectivities and orientations and extramembranous loop sizes and dispositions.

Validity of the model

The model was principally constructed as an aid to future protein-engineering studies and it is important to examine its validity as new information becomes available. Thus, the cysteine-labelling data and the presence and importance of the single disulphide bond at the extracellular face of the protein have been supported by chemical [6] and mutagenesis studies [7]. So too has the elimination of Asp-83, the previous favourite, as a candidate for the important counterion for the protonated Schiff's base [8]. But perhaps the most instructive analysis, came from further model-building studies with the red and green human cone pigments [9]. These proteins are identical, but for 15 residues, and it is not unreasonable to suppose that many of these substituents are associated in some way with the only known functional difference between the two proteins — namely their absorbance maxima. Fig. 1 illustrates that the great majority of the sequence differences occur at residues that line the putative binding pocket of the chromophore. Some of these changes seem to alter the 'polar balance' in this region. Even more dramatic, has been the evidence from a pigment, isolated from a colour-blind individual, whose *N*-terminal residues come from the 'red' visual receptor and the *C*-terminal region from the 'green' pigment [10]. Since the resulting protein exhibits the complete absorbance characteristics of the 'green' receptor, it is reasonable to suppose that these are largely governed by residues 214, 217, 259, 261, 263, 269 and 293, most of which project into the putative chromophore binding-pocket. Data such as these, taken together, lend some credence to the protein chemistry and model-building approach used to generate this working structure for rhodopsin.

Extrapolation to Other Receptors

Part of the usefulness of this three-dimensional representation is its transfer to other members of the family with different ligand and G-protein specificities. Since the degree of identity between the primary structures of the various subgroups can be below 20%, the important sequence alignments necessary for model building can be problematical. To facilitate such alignments and to further our analysis of the intramembranous domain of the receptors, we have developed a series of seven discriminators, one for each transmembrane segment (T. K. Attwood, E. E. Eliopoulos & J. B. C. Findlay, unpublished work). These discriminators were first constructed for the rhodopsin group and were compiled manually in the form of a two-dimensional matrix in which each position in the transmembrane 'motif' was given an index weighted on the basis of the type of amino acid found at that position across the whole range of rhodopsin sequences. As shown in Fig. 2, these discriminators were able accurately to identify the transmembrane regions in the other distant receptor groups, and their individual diagnostic performance was attested by the fact that each had a unique solution, despite the overall hydrophobic character of all seven regions. The quality of the match is indicative of the high level of residue character conservation in these regions, as

HELIX I				HELIX II			
I†	I-1	C	C-1	I	I-1	C	C-1
ASN 55*		LEU 57			LEU 79	LEU 76	
LEU 68		LEU 59			ASP 83	ASN 78	
HELIX III				**HELIX IV**			
I	I-1	C	C-1	I	I-1	C	C-1
CYS 110		LEU 131		TRP 175	TRP 161	VAL 157	ALA 153
ARG 135		GLU 134			PRO 171		
		TYR 136					
		VAL 139					
HELIX V				**HELIX VI**			
I	I-1	C	C-1	I	I-1	C	C-1
PRO 215		VAL 218		GLU 247	TRP 265	LYS 248	VAL 254
TYR 223		ILE 219		PRO 267		VAL 250	
		LEU 226				PHE 261	
						TYR 268	
HELIX VII				**LOOPS**			
I	I-1	C	C-1	I	I-1	C	C-1
PRO 303		PHE 313	ARG 314	ASN 73			PRO 142
TYR 306				GLY 106			
				CYS 187			

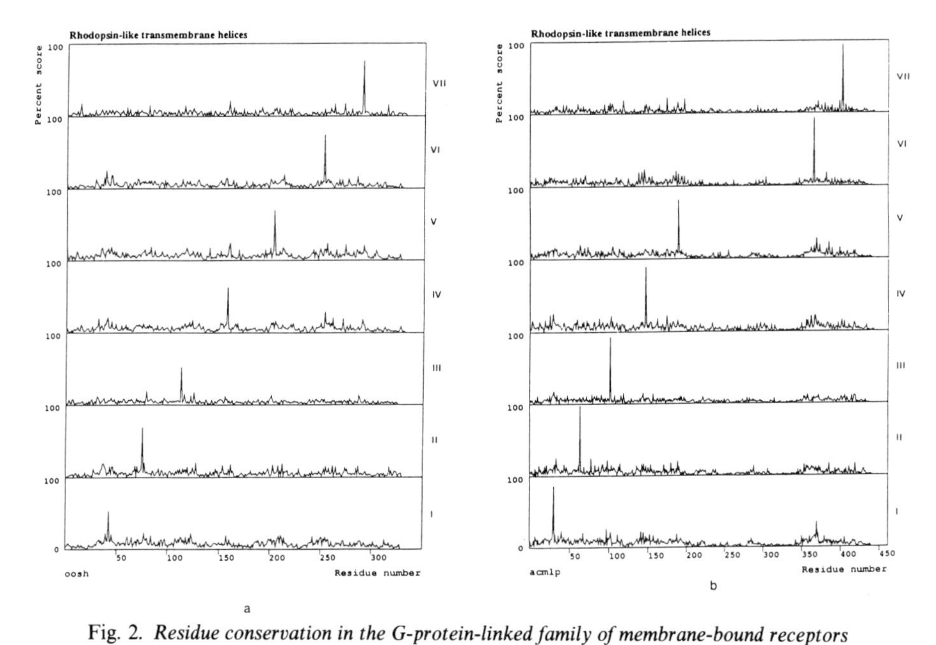

Fig. 2. *Residue conservation in the G-protein-linked family of membrane-bound receptors*

Above: I, identity; I-1, identity except for 1 receptor; C, conserved side-chain character and C-1, conserved character except for 1 receptor. *Numbering as for ovine rhodopsin. Below, discriminator scans of (*a*) ovine rhodopsin and (*b*) porcine muscarinic receptor M1, high-lighting seven transmembrane motifs in each sequence.

distinct from sequence conservation, which may imply that certain structural aspects are also conserved.

This approach has allowed us confidently to align a large number of sequences and to assess, in the context of the model, what may be important structural and functional features for the family as a whole. The picture that emerges is very revealing and indicates a distinct pattern of residue conservation: virtually all conserved features appear in the transmembrane and cytoplasmic regions of the molecule, i.e. not at the extracellular face. From this we infer that conservation in the transmembrane regions will have a structural role, perhaps maintaining the integrity of helix packing, or contributing to the fairly rigid framework defining the limits of the binding pocket: we would mention, in this context, the four prolines that appear to surround the putative ligand-binding site and the three aromatic residues on helix VI that present, in effect, an aromatic back-drop behind the chromophore, sealing a potential gap created by the possible distortion of helix VII, associated with the polar triplet adjacent to the chromophore attachment site (Lys-296). Conversely, the disparity of conserved residues falling at the cytoplasmic and extracellular faces carries a functional inference, i.e. is consistent with a common interaction, at the cytoplasmic end, with G-proteins: the acidic–arginine–aromatic triplet (residues 134–136) at the top of helix III should prove to be a particularly strong candidate for such interactions, others that might be important being Glu-247, Lys-248 and Arg-314.

It was further interesting to note that: (i) conserved polar residues that fell within the transmembrane regions were almost always positioned on internal or interfacial sides of the helices — one interesting group of polar residues could well form an interacting triplet, namely Asn-55 and Asn-302 (or Asp in two *Drosophila* sequences), which may sandwich the almost totally conserved intramembranous negative charge, Asp-83; (ii) all but one of the conserved aromatics were located on external faces, closer inspection of the model revealing that all resided on the helices that house the four prolines — the sole internally oriented aromatic was that associated with the acid–arginine–aromatic triplet on helix III; and (iii) a series of conserved non-polar and aromatic residues, at the cytoplasmic end of the molecule, appear to form a collar, defining, in effect, an apolar interface between the innermembranous and the headgroup portions of the helices in this region. For a full list of the most prominently conserved residues, refer to Fig. 2.

Muscarinic Receptor

From the viewpoint of pharmacology, the adrenergic and muscarinic subgroups of receptors present the most attractive candidates for structure/function analysis. Accordingly, we have constructed a model for the human M1 (muscarinic) receptor using the alignments indicated earlier. Residue-by-residue examination of the transmembrane segments revealed no major inconsistencies; very few polar residues appeared to be exposed to a non-polar environment, a situation also seen in the photosynthetic reaction centre complex [12].

In contrast, the central cavity of the receptor, a position equivalent to the retinal binding-pocket of rhodopsin, appears very polar in character. Hydroxyl-containing residues, serines, threonines and asparagines are particularly evident. In addition, there appears to be a connecting channel linking this site to an equally polar cavity surrounding a negatively charged side-chain (Asp-105) located near to the extracellular face of the receptor. On the basis of the structure, therefore, it is tempting to envisage some form of communica-

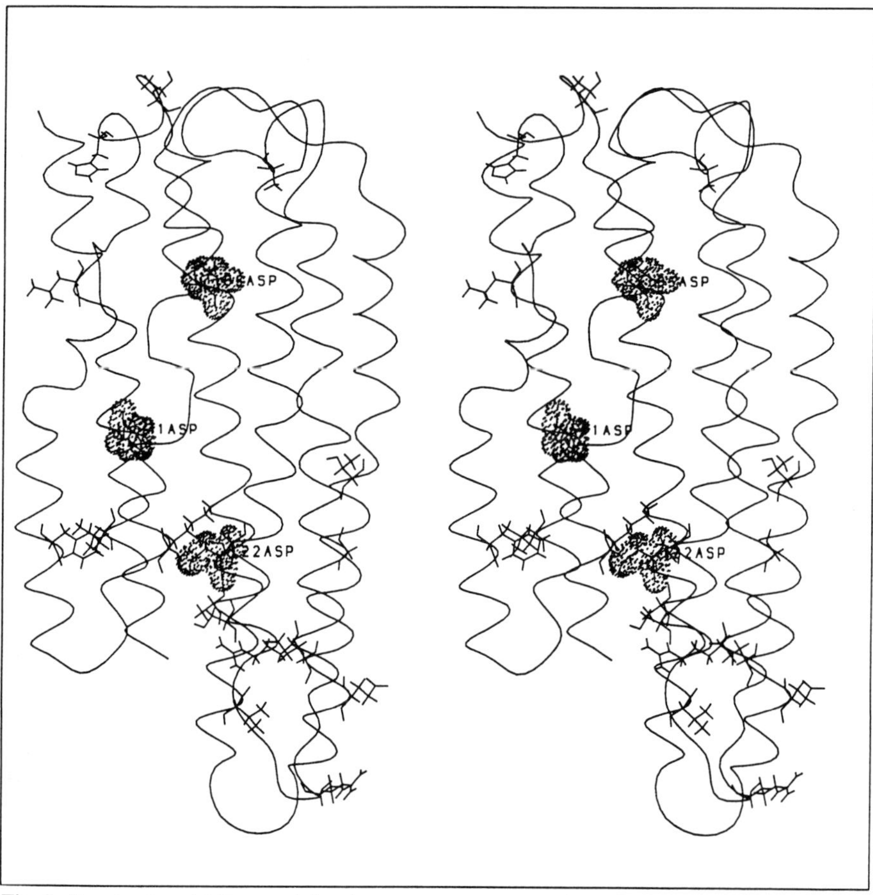

Fig. 3. *Structural model (in stereo) for the muscarinic receptor showing the three aspartic acid residues discussed in the text, together with the positions of those side-chains that vary between the M1, 3 and 5 and M2 and 4 subtypes*

tion route existing between the superficial (105) and buried (71) aspartic acid residues. It has been suggested, for example, that this channel may act as the conduit for the natural ligand and synthetic agonists [13]. Activation of the receptor as a result of ligand or agonist binding may then involve conformational changes which bring into play a third, highly conserved, aspartic acid (122) at the cytoplasmic face of the receptor (this latter residue may have some involvement with G-protein binding). Some support for such a process comes from mutagenic studies. Replacement of the equivalent of Asp (105) in the β-adrenergic receptor almost completely abolishes both antagonist and agonist binding to the protein, while substitution of Asp (71) affects only agonist binding [14]. Mutation of Asp (122) seems to be associated more with G-protein coupling [15].

All the indications are that the conformational changes that link agonist binding to G-protein activation are likely to be substantial, requiring, unlike the photosynthetic reaction centre, a rather 'flexible' structure. Particularly intriguing in this respect is the short stretch of polar residues that follows the attachment point for retinal in rhodopsin (and its equivalent in the other receptors). Such evidence as we have tends to suggest that this region may adopt a non-regular conformation rather than a regular helix. Its involvement with the putative ligand-binding site may point to a degree of participation in the conformational changes which produce activation [1,2].

The most obvious distinguishing feature between agonists and most antagonists is the presence of extended hydrophobic moieties on the latter groups of compounds. It has been suggested that this feature is primarily responsible for conferring antagonist character. It is interesting to note, therefore, a ring of hydrophobic side-chains at the extracellular termini of helices II, III, VI and VII which may provide part of the approaches to the binding site of the receptor. Whether they contribute to antagonist subtype specificity requires a more complex study.

Finally, the muscarinic receptors show internal specificity as to the particular signal transduction pathway affected, be it either adenylate cyclase or PI phospholipase C. It was of interest, therefore, to examine sequence variabilities between the various subtypes (Fig. 3). The picture that emerges is intriguing in that most of the differences are found at or near the cytoplasmic face of the receptor. Thus, in addition to preserving the fundamental property of G-protein binding and activation, the receptors appear to possess individual subtleties, which may allow them to distinguish between different G-proteins. The model does indicate a number of residues which may be responsible for the ability of the receptors to show G-protein specificity, but only experimentation involving mutagenesis will identify the critical side-chains.

Conclusion

As with most integral membrane proteins, the production of crystals suitable for diffraction analysis has not yet been reported, despite considerable efforts round the world. It is unlikely, therefore, that structural information at even low resolution will be available in the near future. High resolution data, of the kind necessary to address the issues raised in this contribution, lie much further away. However, the advent of mutagenic techniques will allow the accumulation of some data, but both the targets for such studies and any interpretation of the observations do depend on some kind of three-dimensional representation of structure. This contribution describes an attempt at such a representation, one which so far has predicted the results surprisingly well. How accurately it reflects the true structure, however, can only be assessed by more detailed analysis.

References

1. Findlay, J. B. C. & Pappin, D. J. C. (1986) *Biochem. J.* **238**, 625–642
2. Findlay, J. B. C., Pappin, D. J. C. & Eliopoulos, E. E. (1988) *Prog. Retinal Res.* **7**, 63–87
3. Pappin, D. J. C., Eliopoulos, E. E., Brett, M. & Findlay, J. B. C. (1984) *Int. J. Biol. Macromol.* **6**, 73–76
4. Henderson, R. & Unwin, P. N. T. (1975) *Nature (London)* **257**, 28–32

5. Davison, M. D. & Findlay, J. B. C. (1986) *Biochem. J.* **236**, 389–395
6. Al-Saleh, S., Gore, M. & Akhtar, M. (1987) *Biochem. J.* **246**, 131–137
7. Karnik, S. S., Sakmar, T. P. & Khorana, H. G. (1988) *Proc. Natl. Acad. Sci. U.S.A.* **85**, 8459–8463
8. Sakmar, T. P., Franke, R. R. & Khorana, H. G. (1989) *Proc. Natl. Acad. Sci. U.S.A.* **86**, 8309–8313
9. Nathans, J., Thomas, D. & Hogness, D. S. (1986) *Science* **232**, 193–202
10. Neitz, J., Neitz, M. & Jacobs, G. N. (1989) *Nature* (*London*) **342**, 679–681
11. Reference deleted
12. Reisenhofer, J., Epp, A., Miki, K., Huber, R. & Michel, H. (1985) *Nature* (*London*) **318**, 618–624
13. Saunders, J. & Freedman, S. B. (1989) in *TIBS, Supplement, Subtypes of Muscarinic Receptors IV* (Levine, R. R. & Birdsall, N. J. M., eds.), pp. 70–75, Elsevier, Amsterdam
14. Strader, C. D., Sigal, I. S., Candelore, M. R. & Dixon, R. A. F. (1988) *J. Biol. Chem.* **263**, 10267–10271
15. Fraser, C. M., Chung, F-Z. & Venter, J. C. (1988) *Proc. Natl. Acad. Sci. U.S.A.* **85**, 5478–5482

Biochem. Soc. Symp. **56**, 9–12
Printed in Great Britain

Regulation of Acetylcholine Receptor Gene Expression by Neural Factors and Electrical Activity During Motor Endplate Formation

J. P. CHANGEUX, P. BENOÎT, A. BESSIS, J. CARTAUD*, A. DEVILLERS-THIÉRY, B. FONTAINE, J. L. GALZI, A. KLARSFELD, R. LAUFER, C. MULLE, H. O. NGHIÊM, M. OSTERLÜND, J. PIETTE and F. REVAH

*Institut Pasteur, Laboratoire de Neurobiologie Moléculaire, 25 rue du Dr Roux, 75724 Paris Cedex 15, France and *Laboratoire de Microscopie Electronique, Institut Jacques Monod, Université Paris 7, 2 place Jussieu, Tour 43, 75251 Paris Cedex 05, France*

The acetylcholine receptor (AChR) light form from fish electric organ and vertebrate neuromuscular junctions is a heterologous pentamer $\alpha_2\beta\gamma\delta$ which contains the ion channel and all the structural elements engaged in the regulation of its opening by acetylcholine.

The amino acids belonging to the acetylcholine-binding site have been identified, on the amino acid sequence of the α-subunit from *Torpedo marmorata*, in its native conformation with a photolabile antagonist (DDF), within three distinct domains which include (Tyr-190-Cys-192–193), (Trp-149) and (Tyr-93) ([1]; J. L. Galzi, F. Revah, D. Black, M. Goeldner, C. Hirth & J. P. Changeux, unpublished work) and are thus exposed to the synaptic cleft. In the presence of acetylcholine, after u.v. irradiation, the non-competitive blocker, [^{3}H]chlorpromazine bound to its high affinity site (present as a unique copy per pentamer and common to all subunits [2]), covalently labels, in the δ-subunit, Ser-262 [3]. The homologous amino acid Ser-254 is labelled on the β-subunit (and in addition Leu-257) [4] as the homologous serines on the α- and γ-subunits pointing to a pseudosymmetrical organization of the high-affinity site for non-competitive blockers along the 5-fold axis of the molecule. These residues belong to the hydrophobic segment termed MII which is highly conserved in the four AChR subunits and is assumed to be a component of the ion channel [3]. This notion is supported by the results of the labelling experiments of Oberthür *et al.* [5] and Hucho *et al.* [6] with another non-competitive blocker and by the subsequent site-directed mutagenesis experiments of Imoto *et al.* [7,8] and Leonard *et al.* [9].

At rest, the receptor is present in a low-affinity 'activatable' conformation; the agonists, some of the non-competitive blockers, Ca^{2+} ions, electric fields, the polypeptide thymopoietin (Goldstein and co-workers [10]) and indirectly via phosphorylation, the neuropeptide calcitonin gene-related peptide (CGRP) [11], stabilize and/or accelerate the fast (100 ms time-scale) and slow (seconds to minutes time-scale) conformational transitions toward a high-affinity 'desensitized' conformation where the ion channel is closed. Such allosteric transitions may serve to model short-term regulation of synapse efficacy at the post-synaptic level (including the Hebb synapse) [12].

In the adult motor endplate, the AChR is strictly localized under the nerve ending, while in the non-innervated myotube, it is distributed all over the surface of the cell. The genesis of this anisotropic distribution involves a sequence of molecular interactions which include regulations at the transcriptional and post-transcriptional levels [13]. *In situ* hybridization with α-subunit probes, containing or not exonic sequences [14,15] disclose high levels of unspliced and mature mRNA in mononucleated myotomal cells and differentiating myotubes. After the entry of the exploratory motor axons, the clusters of grains located outside the endplate decrease in number. In 15-day-old chicks, AChR α-subunit mRNAs become restricted to the subneural 'fundamental' nuclei. Denervation causes a reappearance of unspliced and mature mRNA in extrajunctional areas.

Chronic paralysis of the embryo by Flaxedil interferes with the disappearance of extrajunctional AChR which thus represents an electrical activity-dependent repression of AChR genes. This process has been analysed with a chicken α-subunit genomic probe in primary cultures of chick myotubes. Blocking their spontaneous electrical activity by tetrodotoxin (TTX) causes increases of both precursor and mature α-subunit mRNA levels, while α-actin mRNA levels do not change [16,17]. The entry of Ca^{2+} ions through the sarcolemmal membrane during electrical activity and possibly the activation of protein kinase C contribute to the repression of α-subunit gene transcription [17].

The maintenance and late increase in AChR number at the endplate level requires the intervention of an anterograde signal from neural origin. CGRP, a peptide shown to coexist with acetylcholine in chick spinal cord motoneurons, increases surface AChR and α-subunit unspliced and mature mRNA respectively by 1.5- and 3-fold [18–20]. CGRP increases the cyclic AMP content of myotubes and stimulates membrane-bound adenylate cyclase in the range of concentrations where it enhances AChR α-subunit gene expression [21]. The phorbol ester 12-*O*-tetradecanoylphorbol 13-acetate abolishes the increase of α-subunit mRNA caused by TTX, but not by CGRP, suggesting that distinct second messengers are involved in the regulation of AChR biosynthesis by electrical activity and by CGRP [19].

The data are interpreted in terms of a model [22–24] which assumes that: (i) in the adult muscle fibre, nuclei may exist in different stages of gene expression in subneural and extrajunctional areas; (ii) different second messengers elicited by neural factors or electrical activity regulate the state of transcription of these nuclei via *trans*-acting allosteric proteins binding to *cis*-acting DNA regulatory elements.

To look for such components, the 5′-end and part of the upstream flanking region of the α-subunit gene was isolated and sequenced in the chick [25]. Transcription initiates at the same position in innervated and denervated muscles. TATA and CAAT boxes and a potential *Sp1* binding and Simian virus 40 core enhancer sites are found upstream. This α-subunit promoter, including 850 bp of the 5′ flanking sequence, was inserted into a plasmid vector in front of a chloramphenicol acetyltransferase (CAT) gene. This construct directed high CAT expression in transfected mouse C2.7 myotubes, but not in unfused C2.7 myoblasts or non-myogenic mouse 3T6 cells. In agreement with the results of Wang *et al.* [26], deletion mapping shows that the −110 to −45 segment of this upstream sequence confers developmental control of expression. DNAase I foot-printing and gel retardation assays show that nuclear factors bind to three distinct domains, AR I, II and III, located within the most proximal 110 nt domains. Levels of several of these factors change during fusion of myoblasts into myotubes (AR IIb and III) and as a consequence of denervation (AR IIb and III) [27].

In the mouse, the genes coding for the four subunits have been located [28] on chromosomes 1 (γ- and δ-subunits), 11 (β-subunit) and 2 (α-subunit) [29] and are, in the chick, discoordinately regulated by electrical activity, TTX and CGRP [16].

Finally, multiple post-transcriptional processes involving, in particular the Golgi apparatus [30], proteins from the basal lamina and from the cytoskeleton (the 43 kDa protein among others) [31], contribute to the clustering and stabilization of the AChR in the post-synaptic membrane.

References

1. Dennis, M., Giraudat, J., Kotzyba-Hibert, F., Goeldner, M., Hirth, C., Chang, J. Y., Lazure, C., Chrétien, M. & Changeux, J. P. (1988) *Biochemistry* **27**, 2346–2357
2. Heidmann, T., Oswald, R. & Changeux, J. P. (1983) *Biochemistry* **22**, 3112–3127
3. Giraudat, J., Dennis, M., Heidmann, T., Chang, J. Y. & Changeux, J. P. (1986) *Proc. Natl. Acad. Sci. U.S.A.* **83**, 2719–2723
4. Giraudat, J., Dennis, M., Heidmann, T., Haumont, P. T., Lederer, F. & Changeux, J. P. (1987) *Biochemistry* **26**, 2410–2418
5. Oberthür, W., Muhn, P., Baumann, H., Lottspeich, F., Wittman-Liebold, B. & Hucho, F. L. (1986) *EMBO J.* **5**, 1815–1819
6. Hucho, F. L., Oberthür, W. & Lottspeich, F. (1986) *FEBS Lett.* **205**, 137–142
7. Imoto, K., Methfessel, C., Sakmann, B., Mishina, M., Mori, Y., Konno, T., Fukuda, K., Kurasaki, M., Bujo, H., Fujita, Y. & Numa, S. (1986) *Nature (London)* **324**, 670–674
8. Imoto, K., Busch, C., Sakmann, B., Mishina, M., Konno, T., Nakai, J., Bujo, H., Mori, Y., Fukuda, K. & Numan, S. (1988) *Nature (London)* **335**, 645–648
9. Leonard, R. J., Labarca, C. G., Charnet, P., Davidson, N. & Lester, H. A. (1988) *Science* **242**, 1578–1581
10. Revah, F., Mulle, C., Pinset, C., Audhya, T., Goldstein, G. & Changeux, J. P. (1987) *Proc. Natl. Acad. Sci. U.S.A.* **84**, 3477–3481
11. Mulle, C., Benoit, P., Pinset, C., Roa, M. & Changeux, J. P. (1988) *Proc. Natl. Acad. Sci. U.S.A.* **85**, 5278–5732
12. Heidmann, T. & Changeux, J. P. (1982) *C.R. Acad. Sci. Paris* **295**, 665–670
13. Laufer, R. & Changeux, J. P. (1989) *J. Biol. Chem.* **264**, 2683–2689
14. Fontaine, B., Sassoon, D., Buckingham, M. & Changeux, J. P. (1988) *EMBO J.* **7**, 603–609
15. Fontaine, B. & Changeux, J. P. (1989) *J. Cell Biol.* **108**, 1025–1037
16. Klarsfeld, A. & Changeux, J. P. (1985) *Proc. Natl. Acad. Sci. U.S.A.* **82**, 4558–4562
17. Klarsfeld, A., Laufer, R., Fontaine, B., Devillers-Thiéry, A., Dubreuil, C. & Changeux, J. P. (1989) *Neuron* **2**, 1229–1236
18. Fontaine, B., Klarsfeld, A., Hökfelt, T. & Changeux, J. P. (1986) *Neurosci. Lett.* **71**, 59–65
19. Fontaine, B., Klarsfeld, A. & Changeux, J. P. (1987) *J. Cell Biol.* **105**, 1337–1342
20. Osterlünd, M., Fontaine, B., Devillers-Thiéry, A., Geoffroy, B. & Changeux, J. P. (1989) *Neuroscience* **32**, 279–287
21. Laufer, R. & Changeux, J. P. (1987) *EMBO J.* **6**, 901–906
22. Changeux, J. P., Klarsfeld, A. & Heidmann, T. (1987) in *The Cellular and Molecular Basis of Learning* (Changeux, J. P. & Konishi, M., ed.), pp. 31–83, Wiley, London

23. Changeux, J. P., Devillers-Thiéry, A., Giraudat, J., Dennis, M., Heidmann, T., Revah, F., Mulle, C., Heidmann, O., Klarsfeld, A., Fontaine, B., Laufer, R., Nghiêm, H. O., Kordeli, E. & Cartaud, J. (1987) in *Strategy and Prospects in Neuroscience* (Hayaishi, O., ed.), pp. 29–76, Japan Scientific Societies Press, Tokyo; VNU Science Press, BV Utrecht
24. Laufer, R. & Changeux, J. P. (1989) *Mol. Neurobiol.* **3**, 1–53
25. Klarsfeld, A., Daubas, P., Bourrachot, B. & Changeux, J. P. (1987) *Mol. Cell Biol.* **7**, 951–955
26. Wang, Y., Xu, H. P., Wang, X. M., Ballivet, M. & Schmidt, J. (1988) *Neuron* **1**, 527–534
27. Piette, J., Klarsfeld, A. & Changeux, J. P. (1989) *EMBO J.* **8**, 687–694
28. Heidmann, O., Buonanno, A., Geoffroy, B., Robert, B., Guénet, J. L., Merlie, J. P. & Changeux, J. P. (1986) *Science* **234**, 866–868
29. Siracusa, L. D., Silan, C. M., Justice, M. J., Jenkins, N. A. & Copeland, N. G. (1989)
30. Jasmin, B. J., Cartaud, J., Bornens, M. & Changeux, J. P. (1989) *Proc. Natl. Acad. Sci. U.S.A.* **86**, 7218–7222
31. Kordeli, E., Cartaud, J., Nghiêm, H. O., Devillers-Thiéry, A. & Changeux, J. P. (1989) *J. Cell Biol.* **108**, 127–139

Biochem. Soc. Symp. **56**, 13–19
Printed in Great Britain

Mutational Analysis of the Epidermal Growth Factor-Receptor Kinase

JOSEPH SCHLESSINGER

Rorer Biotechnology Inc., 680 Allendale Road, King of Prussia, PA 19406, U.S.A.

Synopsis

The biological responses of epidermal growth factor (EGF) are mediated by a surface receptor denoted as the EGF receptor. The EGF receptor possesses intrinsic protein tyrosine kinase activity which is essential for signal transduction. Recent evidence shows that EGF receptor phosphorylates several substances including: phospholipase C-γ and the GTPase-activating protein (GAP). Moreover, these proteins become associated with the activated receptor in an immunocomplex. Autophosphorylation of the EGF receptor appears to be required for the association with phospholipase C-γ. Mutational analysis indicates that the intrinsic autophosphorylation sites compete with exogenous substrates for the substrate-binding site in the kinase domain.

The ligand-binding site for EGF was analysed using a chimeric receptor approach. Subdomains of the extracellular ligand-binding region of the chicken EGF receptor, which binds EGF with low affinity, were replaced by corresponding regions of the human EGF receptor, which binds EGF with high affinity. On the basis of this analysis, it is concluded that subdomain III of the extracellular domain of the EGF receptor is a major ligand-binding domain. Together, domain I and domain III are able to reconstitute nearly all interactions which bring about high-affinity binding.

Growth factor receptors with protein tyrosine kinase (PTK) activity could be envisioned as membrane-associated allosteric enzymes. Unlike water-soluble allosteric enzymes, the configuration of the growth factor receptors dictates that the ligand-binding domain and PTK activity of the receptor molecules are separated by the plasma membrane. Therefore, ligand-induced signal must cross the membrane barrier to activate the PTK function. In principle, ligand control of receptor kinase activity could involve either an intramolecular or an intermolecular mechanism (reviewed in [1,2]). Both models based on either intramolecular or intermolecular mechanisms assume that ligand binding induces a conformational change in the extracellular domain and that this change leads to the activation of the cytoplasmic domain by an allosteric transition. According to the model based on the intramolecular mechanism, the conformational change induced by the ligand in the extracellular domain causes a vertical perturbation in the transmembrane region which is followed by a conformational change in the cytoplasmic domain and enhancement of the PTK activity [1,2]. However, according to the intermolecular mechanism, the conformational change in the extracellular domain causes receptor oligomerization. This facilitates interaction between neighbouring cytoplasmic domains and enhancement of PTK activity by subunit interaction [1,2]. Receptor oligomerization, according to this model, facilitates the transfer of conformational change from the extracellular domain to the cytoplasmic domain without altering the positioning of residues of the transmembrane region. Receptor oligomerization appears to be a general phenomenon among growth factor receptors; it was detected in living cells, in membranes and in solubilized and purified receptor preparations [1–7]. Moreover, it was shown that the oligomeric state possesses elevated PTK activity and enhanced ligand-binding affinity [5,6].

Growth factor receptors with PTK activity catalyse the phosphorylation of both exogenous substrates and of multiple intrinsic sites (reviewed in [1,8]). After ligand binding and activation, the insulin receptor undergoes rapid autophosphorylation [9]. This increases the $V_{max.}$ of the kinase activity, maintaining the PTK in the activated state even in the absence of ligand binding. The intrinsic autophosphorylation sites of the EGF receptor are clustered at the *C*-terminal tail of the receptor and they have a different regulatory role. The autophosphorylation sites of EGF receptors compete with exogenous substrates for the substrate-binding region of the PTK domain. Hence, autophosphorylation releases an internal constraint and renders the PTK of EGF receptor more accessible towards exogenous substrates [10,11]. Recent evidence suggests that autophosphorylation of the solubilized EGF receptor is mediated by cross-phosphorylation [12,13] suggesting that receptor oligomerization provides the mechanism for receptor activation and also sets the stage for autophosphorylation.

Pathways of Signal Transduction

The addition of growth factors to cells leads to rapid enhancement of an array of early responses. These include stimulation of Na^+/H^+ exchange, Ca^{2+} influx, activation of phospholipase C and stimulation of glucose and amino acid transport. The stimulation of phospholipase C leads to the generation of phosphoinositol metabolites, such as inositol 1,4,5-trisphosphate (Ins[1,4,5]P_3) which causes the release of Ca^{2+} from intracellular compartments and the generation of the natural activator of protein kinase C, *sn*-1,2-diacylglycerol (DAG). Numerous intrinsic cellular substrates are directly phosphorylated on tyrosine residues by the PTK activity of growth factor receptors. Growth factors also stimulate the phosphorylation of multiple substrates on serine and threonine residues. Some of these substrates are phosphorylated by protein kinase C, by S6 kinase or by other unidentified Ser/Thr kinases. It is assumed that the phosphorylation of cellular substrates, together with alterations in ionic content of the cell, provides an internal stimulus for cell growth. However, the chain of events which is initiated by tyrosine phosphorylation is not yet established, and more work is required to identify the relevant substrates of PTK and their role in signal transduction.

Receptor Transmodulation

The binding of platelet-derived growth factor (PDGF) or bombesin to their respective receptors leads to the abolition of the high-affinity binding of EGF towards its own receptor [1,14,15]. Similarly, the phorbol ester (12-*O*-tetradecanoylphorbol 13-acetate; PMA) abolishes the high-affinity binding of EGF receptor and reduces its PTK activity [15–19]. PMA activates protein kinase C, which, in turn, phosphorylates the EGF receptor on serine and threonine residues, including Thr-654 [18,19]. Since both bombesin and PDGF are able to stimulate the inositol phospholipid signalling pathway, it was proposed that their effect on EGF binding affinity is mediated by protein kinase C. Hence, protein kinase C may act as a messenger for receptor transmodulation and this could provide a negative feedback mechanism for the control of receptor activity [20,21].

Another example of heterologous receptor control relates to the mechanism of EGF-induced phosphorylation of HER2/*neu* by EGF. It was shown that the binding of EGF to its own receptor triggers rapid tyrosine phosphorylation of the *neu* protein [22,23]. This cross-phosphorylation could be mediated by the formation of mixed oligomers containing activated EGF receptor and activated *neu* proteins which are stabilized by the binding of EGF to the hetero-oligomeric complex. The effect of *neu*-phosphorylation on its enzymic or biological properties is not yet established.

Structure–Function Analysis of the EGF Receptor Utilizing Site-Directed Mutagenesis

The extracellular domain

The extracellular domain of the insulin receptor (α-subunit) is related to the extracellular domain of the EGF receptor as both domains contain highly homologous cysteine-rich regions (reviewed in [8]). The extracellular domains of the PDGF receptor, CSF_1 receptor, FGF-receptors and other members of this receptor family evolved from a different ancestral receptor gene which contains three or five related 'immunoglobulin like' repeats [8,24–27]. Affinity labelling experiments with ^{125}I-EGF indicated that the region flanked by the two cysteine-rich domains denoted domain III serves as a major ligand-binding domain of the EGF receptor [28]. These results are consistent with a functional analysis of the ligand-binding domain utilizing chimeras between human and chicken EGF receptors [29]. This approach is based on the fact that the chicken EGF receptor binds EGF with 100-fold lower affinity than the human EGF receptor [30]. It was shown that a chicken EGF receptor containing domain III of the human EGF receptor behaves like the human EGF receptor with respect to EGF binding and biological responsiveness. Monoclonal antibodies which block ligand binding recognize specifically domain III of the human EGF receptors. A similar approach was used to demonstrate that domain I also contributes part of the forces which define ligand-binding specificity of EGF towards its receptor (I. Lax, F. Bellot, A. Honegger, A. Schmidt, A. Ullrich, D. Givol & J. Schlessinger, unpublished work). On the basis of these analyses, the internal homology in the primary structures of EGF receptor [32], and the fact that other growth factor receptors contain similar domains, we propose a four-domain model for the organization of the extracellular portion of the EGF receptor. Each of the four domains may fold independently to form a structure with some specific function. Domain III which is flanked by the two cysteine-rich domains together with domain I contribute to most of the forces which define EGF binding specificity [32]. Thus it is proposed that EGF is able to interact with domains III and I, and, therefore, the ligand-binding region may lie in the cleft formed between these two domains. Such a configuration for the ligand-binding region is common in many allosteric enzymes, where ligand binding alters the interaction between neighbouring subunits thus allowing transfer of an allosteric conformational transition. According to this model, the two cysteine-rich domains are in contact with each other and are close to the plasma membrane.

Transmembrane region

The *neu* proto-oncogene product becomes activated upon substitution of a single valine residue by a glutamic acid residue in its transmembrane region [33–35]. This mutation leads to the activation of the PTK function [35], which is essential for the transforming potential of *neu*. However, the integrity of the transmembrane region is not essential for signal transduction. It was shown that EGF is able to enhance the kinase activity and to stimulate various responses in cells expressing mutant EGF receptors with altered transmembrane regions [36]. The transmembrane regions of two mutant receptors were either shortened or extended by three hydrophobic amino acids, and in the other two mutant receptors, hydrophobic amino acids were substituted by charged residues [36]. Experiments with chimeric receptors including portions of the EGF receptor HER2/*neu*, PDGF receptor or insulin receptor indicated that the origin of the transmembrane region did not alter the signalling capacity of the chimeric receptor molecules [37–39]. The origin of the extracellular domain determined ligand specificity, while the origin of the cytoplasmic domain determined biological responsiveness (K. Seedorf, S. Felder, L. Mosthaf, J. Schlessinger & A. Ullrich, unpublished work; [38–41]). The origin of the transmembrane region did not influence the capacity of the two chimeric regions to activate each other. Hence, the transmembrane region has probably a passive role in signal transduction *per se*, further supporting an allosteric oligomerization mechanism, rather than a mechanism based on intramolecular receptor activation.

Juxtamembrane domain

The juxtamembrane domain of the EGF receptor appears to play an important role in the process of receptor transmodulation. Protein kinase C-induced phosphorylation of the EGF receptor appears to be essential for the abolition of the high binding by PMA [15–19]. However, phosphorylation of Thr-654 does not play a role in the regulation of receptor affinity [21], as was proposed by several investigators [18–20]. Phosphorylation of Thr-654 does, however, provide a negative control mechanism for EGF-induced mitogenesis [21]. The phosphorylation of unidentified serine or threonine residues, other than Thr-654, is probably required for regulation of receptor affinity and PTK activity mediated by PMA. Interestingly, PMA is also able to modulate the binding affinity of EGF to a chimeric receptor molecule composed of the extracellular domain of the EGF receptor fused to the cytoplasmic domain of HER2/*neu*, suggesting that HER2/*neu* is also controlled by similar protein kinase C-mediated heterologous regulation [37].

PTK domain

All PTKs contain a consensus lysine residue and a consensus sequence, GlyXaaGlyXaaXaaGlyXaa, 15 residues upstream to the lysine residue. These residues function as part of the ATP-binding site of the PTK domains of the receptor molecules [1,8,42]. Replacement of the consensus lysine residues of the ATP-binding site of the EGF receptor, insuline receptor, PDGF receptor, and other receptors with PTK activities, completely abolished their kinase activities both *in vitro* and in living cells [43–49]. Additional receptor mutants generated to explore the role of PTK include linker insertion mutants in different positions of the PTK region [50–51] and double-mutated EGF receptor in which both Lys-721 and Thr-654 were mutated [46,48]. While the kinase activity of the various receptors is dispensable for their expression and display on the cell surface, it is indispensable for signal transduction, activation of early and delayed responses, mitogenesis and transformation [43–52]. Although normal in its binding characteristics, the kinase-negative mutant of EGF receptor was unable to stimulate Ca^{2+} influx, inositol phosphate formation, Na^{+}/H^{+} exchange, c-*fos* and c-*myc* expression, S6 ribosomal protein phosphorylation, DNA synthesis and transformation. The PTK activity of the EGF receptor is also essential for normal receptor trafficking. The kinase activity is required for specifically targeting the EGF receptor for degradation, probably by phosphorylating unknown substrate(s) involved in receptor routeing [43]. It appears that signal transduction, mitogenesis and normal receptor traffic depend on a functional protein kinase, suggesting that these processes are regulated by tyrosine phosphorylation of cellular substrates. The inability of the kinase-negative mutant to enhance activation of inositol phosphate formation and calcium mobilization suggests a linkage between the inositol phospholipid signalling pathway and tyrosine kinase activity [45]. It is possible that key regulatory proteins in the inositol phospholipid signalling pathway, such as Gp protein or phospholipase C, are substrates of protein tyrosine kinases, and that tyrosine phosphorylation regulates their activities enabling direct coupling between receptors with PTK activity and the inositol phospholipid signalling pathways.

The C-terminal tail

The *C*-terminal tails of growth factor receptors with PTK activity are the most divergent of the receptors [8]. Several autophosphorylation sites were mapped in this region for the EGF receptor [53,54]. These tyrosine residues are conserved in the *C*-terminal tails of the subclass of each receptor family [8]. The biological role of the *C*-terminal tail and autophosphorylation sites were explored using a series of deletion mutants and mutant receptors in which individual autophosphorylation sites were replaced by phenylalanine residues [10,11,55]. These mutant receptors have similar enzymic properties to wild-type EGF receptors expressed in the same background [10,11]. Autophosphorylation of the EGF receptor does not affect the $V_{max.}$ of the receptor kinase and the apparent K_m of peptide substrates is somewhat lowered upon autophosphorylation. This result is consistent with a mechanism in which exogenous substrates and the intrinsic autophosphorylation sites compete for the active site of the PTK [10,11]. Upon autophosphorylation or deletion of the intrinsic sites, the substrate-binding sites become more accessible for exogenous substrates. The *C*-terminal tail of the receptor may possess enough length and flexibility to interact with the substrate-binding sites of the PTK region and thus modulate its capacity to interact with exogenous substrates. Indeed, the dose–response curves for DNA synthesis of cells expressing mutant EGF receptors which altered autophosphorylation sites were slightly shifted to lower concentrations of EGF, rendering the cells mitogenically responsive to lower doses of EGF than cells expressing wild-type receptor at similar expression levels [11]. Deletions in the *C*-terminal tail have a similar effect on transformation, they potentiate oncogenic capacity and increase host range, but do not provide a major oncogenic lesion [56].

References

1. Schlessinger, J. (1986) *J. Cell. Biol.* **103**, 2067–2072
2. Schlessinger, J. (1988) *Trends Biochem. Sci.* **13**, 443–447
3. Yarden, Y. & Schlessinger, J. (1985) *Growth Factors in Biology and Medicine*, pp. 23–45, Pitman, London
4. Yarden, Y. & Schlessinger, J. (1987) *Biochemistry* **26**, 1434–1442
5. Yarden, Y. & Schlessinger, J. (1987) *Biochemistry* **26**, 1443–1451
6. Boni-Schnetzler, M. & Pilch, F. F. (1987) *Proc. Natl. Acad. Sci. U.S.A.* **84**, 7832–7836
7. Cochet, C., Kashles, O., Chambaz, E. M., Borrello, I., King, C. R. & Schlessinger, J. (1988) *J. Biol. Chem.* **263**, 3290–3295
8. Yarden, Y. & Ullrich, A. (1988) *Annu. Rev. Biochem.* **57**, 443–478
9. Rosen, O. M., Herrera, R., Olowe, Y., Petruzzelli, L. M. & Cobb, M. H. (1983) *Proc. Natl. Acad. Sci. U.S.A.* **80**, 3237–3240
10. Honegger, A. M., Dull, T. J., Szapary, D., Komoriya, A., Kris, R., Ullrich, A. & Schlessinger, J. (1988) *EMBO J.* **7**, 3053–3060
11. Honegger, A. M., Dull, T. J., Bellot, F., Van Obberghen, E., Szapary, D., Schmidt, A., Ullrich, A. & Schlessinger, J. (1988) *EMBO J.* **7**, 3045–3052
12. Honegger, A. M., Kris, R. M., Ullrich, A. & Schlessinger, J. (1989) *Proc. Natl. Acad. Sci. U.S.A.* **86**, 925–929
13. Honegger, A. M., Schmidt, A., Ullrich, A. & Schlessinger, J. (1990) *Mol. Cell. Biol.* in the press
14. Wrann, M., Fox, C. F. & Ross, R. (1980) *Science* **210**, 1363–1364
15. King, A. C. & Cuatrecasas, P. (1982) *J. Biol. Chem.* **257**, 3053–3060
16. Cochet, C., Gill, G., Meisenhelder, J., Cooper, J. A. & Hunter, T. (1984) *J. Biol. Chem.* **259**, 2553–2558
17. Iwashita, S. & Fox, C. F. (1984) *J. Biol. Chem.* **259**, 2559–2567
18. Hunter, T., Ling, N. & Cooper, N. A. (1984) *Nature* (*London*) **314**, 480–483
19. Davis, R. J. & Czech, M. P. (1985) *Proc. Natl. Acad. Sci. U.S.A.* **82**, 1974–1978
20. Lin, C.-R., Chen, S.-W., Lazar, C. S., Carpenter, D. C., Gill, G. N., Evans, R. N. & Rosenfeld, M. G. (1986) *Cell* (*Cambridge, Mass.*) **44**, 839–848

21. Livneh, E., Dull, T. J., Prywes, R., Ullrich, A. & Schlessinger, J. (1988) *Mol. Cell. Biol.* **8**, 2302–2308
22. Stern, D. F. & Kamps, M. P. (1988) *EMBO J.* **7**, 995–1001
23. King, C. R., Borrello, O., Bellot, F., Comoglio, P. & Schlessinger, J. (1988) *EMBO J.* **7**, 1647–1651
24. Williams, L. T. (1989) *Science* **243**, 1564–1570
25. Lee, P. L., Johnson, D. E., Cousens, L. S., Fried, V. A. & Williams, L. T. (1989) *Science* **245**, 57–60
26. Ruta, M., Howk, R., Ricca, G., Drohan, W., Zabelshansky, M., Laureys, G., Barton, D. E., Francke, U., Schlessinger, J. & Givol, D. (1988) *Oncogene* **3**, 9–15
27. Ruta, M., Burgess, W., Givol, D., Epstein, J., Neiger, N., Kaplow, J., Crumley, G., Dionne, C., Jaye, M. & Schlessinger, J. (1989) *Proc. Natl. Acad. Sci. U.S.A.* **86**, 8722–8726
28. Lax, I., Burgess, W. H., Bellot, F., Ullrich, A., Schlessinger, J. & Givol, D. (1988) *Mol. Cell. Biol.* **8**, 1831–1834
29. Lax, I., Bellot, F., Howk, R., Ullrich, A., Givol, D. & Schlessinger, J. (1989) *EMBO J.* **8**, 421–427
30. Lax, I., Johnson, A., Howk, R., Sap, J., Bellot, F., Winkler, M., Ullrich, A., Vennstrom, B., Schlessinger, J. & Givol, D. (1988) *Mol. Cell. Biol.* **8**, 1970–1978
31. Reference deleted
32. Lax, I., Bellot, F., Honegger, A., Schmidt, A., Ullrich, A., Givol, D. & Schlessinger, J. (1990) *Cell Regul.* in the press
33. Bargman, C. I., Hung, M. C. & Weinberg, R. A. (1986) *Cell (Cambridge, Mass.)* **45**, 649–657
34. Coussens, L., Yang-Fend, T. L., Liao, Y. C., Chen, E., Gray, A., McGrath, J., Seeburg, P. H., Libermann, T. A., Schlessinger, J., Francke, Y., Levinson, A. & Ullrich, A. (1985) *Science* **230**, 1132–1139
35. Stern, D. F., Kamps, M. P. & Lao, H. (1988) *Mol. Cell. Biol.* **8**, 3969–3973
36. Kashles, O., Szapary, D., Bellot, F., Ullrich, A., Schlessinger, J. & Schmidt, A. (1988) *Proc. Natl. Acad. Sci. U.S.A.* **85**, 9567–9571
37. Lee, J., Dull, T. J., Lax, I., Schlessinger, J. & Ullrich, A. (1989) *EMBO J.* **8**, 167–173
38. Riedel, H., Dull, T. J., Schlessinger, J. & Ullrich, A. (1986) *Nature (London)* **324**, 68–70
39. Riedel, H., Massoglia, S. L., Schlessinger, J. & Ullrich, A. (1989) *EMBO J.* **8**, 2943–2954
40. Reference deleted
41. Lammers, R., Gray, A., Schlessinger, J. & Ullrich, A. (1989) *EMBO J.* **8**, 1369–1375
42. Hanks, S. K., Quinn, A. M. & Hunter, T. (1988) *Science* **241**, 45–52
43. Honegger, A. M., Dull, T. J., Felder, S., Van Obberghen, E., Bellot, F., Szapary, D., Schmidt, A., Ullrich, A. & Schlessinger, J. (1987) *Cell (Cambridge, Mass.)* **51**, 199–209
44. Honegger, A. M., Szapary, D., Schmidt, A., Lyall, R., Van Obberghen, E., Dull, T. J., Ullrich, A. & Schlessinger, J. (1987) *Mol. Cell. Biol.* **7**, 4567–4571
45. Moolenaar, W. H., Bierman, A. J., Tilly, B. C., Verlaan, I., Defize, L. H. K., Honegger, A. M., Ullrich, A. & Schlessinger, J. (1988) *EMBO J.* **7**, 707–710
46. Chen, S. W., Lazar, S. C., Poenie, M., Tsien, R. Y., Gill, G. & Rosenfeld, M. G. (1987) *Nature (London)* **328**, 820–823
47. Chou, D. K., Dull, T. J., Russell, D. S., Gherzi, R., Lebwohl, D., Ullrich, A. & Rosen, O. M. (1987) *J. Biol. Chem.* **262**, 1842–1847
48. Glenney, J. R., Chen, W. S., Lazar, C. S., Walton, G. M., Zokas, M. R. & Gill, G. N. (1988) *Cell (Cambridge, Mass.)* **52**, 675–684
49. McClain, D. A., Maegawa, H., Lee, J., Dull, T. J., Ullrich, A. & Olefsky, J. M. (1987) *J. Biol. Chem.* **262**, 14663–14671
50. Prywes, R., Livneh, E., Ullrich, A. & Schlessinger, J. (1986) *EMBO J.* **5**, 2179–2190
51. Livneh, E., Reiss, N., Berent, E., Ullrich, A. & Schlessinger, J. (1987) *EMBO J.* **6**, 2669–2676
52. Livneh, E., Prywes, R., Kashles, O., Reiss, N., Sasson, I., Mory, Y., Ullrich, A. & Schlessinger, J. (1986) *J. Biol. Chem.* **261**, 12490–12497
53. Downward, J., Parker, P. & Waterfield, M. D. (1984) *Nature (London)* **311**, 483–485
54. Margolis, B., Lax, I., Kris, R., Dombalagian, M., Honegger, A. M., Howk, R., Givol, D., Ullrich, A. & Schlessinger, J. (1989) *J. Biol. Chem.* **264**, 10667–10671
55. Bertics, P. J., Chen, W. S., Hubler, L., Lazar, C. S., Rosenfeld, M. G. & Gill, G. N. (1988) *J. Biol. Chem.* **263**, 3610–3617
56. Khazaie, K., Dull, T., Graf, T., Schlessinger, J., Ullrich, A. & Vennstrom, B. (1988) *EMBO J.* **7**, 3061–3071

Biochem. Soc. Symp. **56**, 21–34
Printed in Great Britain

Specificity of Interactions of Receptors and Effectors with GTP-Binding Proteins in Native Membranes

GRAEME MILLIGAN*†, IAN MULLANEY* and FERGUS R. McKENZIE*

*Molecular Pharmacology Group, Departments of *Biochemistry and †Pharmacology, University of Glasgow, Glasgow G12 8QQ, Scotland, U.K.*

Summary

Individual G-proteins are highly similar in primary sequence. It is thus pertinent to ask what degree of specificity of interaction each of these display with the various receptors and effector systems. Many of the identified G-proteins are co-expressed in a single tissue or cell. As the extreme *C*-terminus of the α-subunit of each G-protein appears to be a key domain for the interactions of receptors and G-proteins, we have generated a series of G-protein-selective anti-peptide antisera against this region and then have used these antisera to attempt to interfere with receptor–G-protein coupling. With this approach, we have demonstrated that δ-opioid receptor-mediated inhibition of adenylate cyclase in neuroblastoma × glioma (NG108-15) cell membranes is transduced specifically by G_i2 and in the same cell that α_2-adrenergic inhibition of Ca^{2+} currents is transduced by G_o.

Introduction

The ‘classical’ guanine-nucleotide-binding proteins (G-proteins) are a family of highly homologous, heterotrimeric species which allow communication between cell surface receptors for hormones, growth factors and neurotransmitters, and the effector systems which convert this primary message within the cell into a signal which will generate a physiological response [1,2].

Each of the ‘classical’ mammalian G-proteins consists of distinct α- (39–46 kDa), β- (35–36 kDa) and γ- (8 kDa) subunits. While genetic diversity at the level of both the β- and γ-subunits has been reported [3,4], there is no overwhelming evidence at this stage to indicate that different agglomerations of distinct β/γ-subunits either subserve distinct functions or indeed interact selectively with different α-subunits. While it has been suggested that β/γ-subunits may have primary roles in both hormonal regulation of phospholipase A_2 activity [5] and in receptor-mediated control of the inhibition of adenylate cyclase activity [6], it is generally assumed that it is the α-subunits of the various G-proteins which define the specificity and selectivity of both receptor/G-protein and G-protein/effector interactions. It should be noted, however, that genetic analysis in *Saccharomyces cerevisiae* has recently provided support for the concept that polypeptides corresponding to mammalian β- and γ-subunits, rather than an α-subunit, are the key elements in transducing signals from the mating factor receptor in haploid cells [7]. Mammalian systems are not yet amenable to such genetic analysis. Outwith mammalian systems, G-proteins which are highly homologous to their mammalian counterparts have been detected in birds [8], amphibians [8], various invertebrates [9], yeast [10], *Drosophila* [11] and a slime mould [12]. Further, experiments which have relied on either the binding of analogues of GTP, or upon the use of antisera which identify regions of G-protein α-subunits that appear to have been highly conserved throughout evolution, have provided initial evidence for the expression of G-proteins in both plant [13] and bacterial systems [14].

In addition, a considerable family of small (19–27 kDa) GTP-binding proteins have recently been identified. These include members of the ras, ral, rho and ARF families of gene products. While no evidence indicates that these polypeptides can interact with β/γ subunits, and it is possible that they serve different functions, the potential role of at least some of these proteins as proto-oncogenes has stimulated considerable interest in the concept that they may function as transducer elements analogous to the 'classical' G-proteins. This is particularly so in pathways, such as the hydrolysis of polyphosphoinositides, which have been implicated in mitogenic signalling [15,16].

The Interaction of Receptors and G-Proteins

The majority of receptors which interact with G-proteins, either to alter the intracellular concentration of second messengers [e.g. cyclic AMP, cyclic GMP, inositol trisphosphate (InsP_3), *sn*-1,2-diacylglycerol (DAG)] or to modify directly the activity of certain ion channels, belong to a superfamily of proteins in which a single polypeptide is sufficient to provide all of the functions of the receptor. In each case, these receptor polypeptides have seven stretches of some 22–24 contiguous hydrophobic amino acids. These regions are believed to represent transmembrane-spanning elements. As the *N*-terminal regions of these receptor polypeptides have asparagine residues which are likely sites for *N*-linked glycosylation, then, given the presence of seven transmembrane-spanning elements, the likely disposition of the receptor in the membrane is with an extracellular *N*-terminus and hence an intracellular *C*-terminus. The *C*-terminal region contains a number of serine and threonine residues which represent potential sites for phosphorylation and hence regulation of receptor function. The binding site for the hormone appears to be both provided and defined by the topology of the transmembrane-spanning elements and to be within the plane of the plasma membrane. Recent evidence has, however, provided strong evidence for the direct interaction between G-proteins and a number of receptors which appear to have only a single transmembrane-spanning element. Some of the most elegant data on this subject have been provided by Nishimoto *et al.* [17], by demonstrating an interaction of IGF 2 receptors with G_i2, and by M. del C. Vila, G. Milligan, J. A. Fox, M. L. Standaert & R. V. Farese (unpublished work), who have shown that anti-$G_i\alpha$ antibodies are able to prevent the interaction of insulin receptors with pertussis toxin-sensitive G-proteins in BC3H1 murine myocytes.

Receptor number and the affinity of interaction between receptors and hormones can conveniently be assessed in ligand-binding studies. As the binding affinity of agonists (but not antagonists) to a receptor which interacts with a G-protein is modulated by the coupling or otherwise of these two proteinaceous components, then such ligand-binding studies can be used to assess the degree of this interaction (see below). All of the 'classical' G-proteins appear to function in a similar manner. In the resting state, the G-protein, which is located at the cytoplasmic face of the plasma membrane, exists as the heterotrimeric complex of α-, β- and γ-subunits. In this situation, the nucleotide-binding site on the α-subunit is filled with GDP. Upon activation of receptor by hormone, nucleotide exchange occurs on the G-protein such that GDP is released and is replaced by GTP. The GTP-liganded form of the α-subunit now dissociates from the β/γ complex and is able to interact with the relevant effector system. As signalling systems should, of necessity, be activated for only a limited duration, then a termination mechanism is required to inactivate the G-protein. This is achieved by a GTPase activity which is intrinsic to the G-protein α-subunit. This activity hydrolyses the terminal phosphate of GTP, restoring the bound guanine nucleotide to GDP. With GDP bound, the now inactive α-subunit reassociates with the β/γ complex. This completes the cycle and the G-protein then awaits further receptor stimulation. Inbuilt into such a system is the requirement that receptor activation, and the consequent promotion of guanine nucleotide exchange on the G-protein α-subunit, will automatically increase the rate of the entire cycle if the exchange reaction is the rate-limiting point. As such, receptor activation of any G-protein should be detectable at the level of an increase in GTPase activity [19]. While this can indeed often be measured, assays of this nature have generally been of greatest usefulness in signalling cascades in which the receptor interacts with a G-protein which is a substrate for pertussis toxin (see later).

Identification of G-Proteins

The initial identification of both the stimulatory (G_s) and inhibitory (G_i) G-proteins of the adenylate cyclase cascade was based on the recognition that toxins from either *Vibrio cholerae* (cholera toxin) or *Bordetella pertussis* (pertussis toxin) were able to modify this signalling system. Treatment of cells with cholera toxin produces a maximal and persistent stimulation of adenylate cyclase activity. By the use of $[^{32}P]NAD^+$ as substrate, this was shown to be due to the mono-ADP-ribosylation of a polypeptide(s) which in different systems has been reported to be between 42 and 52 kDa. Isolation of cDNA clones has indicated that these two forms are 44.5 and 46 kDa and are derived from the differential splicing of pre-mRNA produced from a single gene [20].

Pertussis toxin also catalyses the mono-ADP-ribosylation of the α-subunits of certain G-proteins [19]. As this was initially shown to correlate with attenuation of receptor-mediated inhibition of adenylate cyclase, then the 41 kDa polypeptide identified after ADP-ribosylation with pertussis toxin and

[^{32}P]NAD^{+} was defined as G_i [21]. Because the adenylate cyclase cascade is an essentially universal signalling system in mammalian cells, and because pertussis toxin was able to identify what appeared to be a single polypeptide in all tissues, then it was assumed both that any approximately 40 kDa polypeptide identified by pertussis toxin-catalysed ADP-ribosylation was 'G_i' and that any hormone function perturbed by pertussis toxin treatment of cells, or membranes derived from these cells, must be transduced by 'G_i'.

This is now appreciated to be an over-simplification. The site of pertussis toxin-catalysed ADP-ribosylation is a cysteine residue located four amino acids from the *C*-terminus of the α-subunit of G-proteins, which are substrates for this toxin. cDNA cloning studies have, to date, demonstrated the existence of six distinct α-subunits (TD1, TD2, G_i1, G_i2, G_i3 and G_o) which have a cysteine residue in this position [22] and each of these polypeptides is indeed a substrate for pertussis toxin-catalysed ADP-ribosylation. While the tissue distribution of the transducins is limited to photoreceptor-containing tissues, the other pertussis toxin-sensitive G-proteins are widely distributed; indeed, both G_i2 and G_i3 appear to be present in all tissues. Such co-expression of these G-proteins implies that treatment of cells with pertussis toxin cannot be used as a means to discriminate selectively between receptor interactions with any of these proteins. However, SDS/PAGE conditions have been defined such that both autoradiography after pertussis toxin-catalysed ADP-ribosylation of membranes using [^{32}P]NAD^{+} as substrate and immunoblotting with mixtures of well-characterized antisera can be used to detect co-expression of a number of pertussis toxin-sensitive G-proteins in a single cell type [23,24].

The extreme conservation of sequence of the individual G-protein α-subunits between mammalian species strongly implies that each G-protein serves a unique function. With this in mind, we have generated anti-peptide antisera able to identify each of the pertussis toxin-sensitive G-proteins. One series of these antisera has been generated against the extreme *C*-terminal decapeptides of the α-subunits [25–27]. As the site for pertussis toxin-catalysed ADP-ribosylation is within this epitope, then we assumed that antibodies able to identify this region might be used to interfere with interactions between a receptor and its G-protein as this is the functional consequence of pertussis toxin-catalysed ADP-ribosylation.

The model system for many of our studies in this area has been the murine neuroblastoma × rat glioma hybrid cell line, NG108-15. Three distinct polypeptides of 40.5, 40 and 39 kDa can be identified by pertussis toxin-catalysed ADP-ribosylation of membranes of these cells. Immunological studies identify these as the α-subunits of G_i3, G_i2 and G_o [28,29]. These three G-proteins may be derived from the rat glioma, as one of the parental cell lines (rat glioma C6 cells) also expresses each of these proteins (G. Milligan, I. Mullaney & F. R. McKenzie, unpublished work). While this cell line is widely used as a 'neuron-like' model system [30], it should be noted that the major 'G_i-like' G-protein in brain is G_i1. However, we have been unable to detect the expression of either the $G_i1\alpha$ polypeptide or of the corresponding mRNA in these cells [28,29] (Fig. 1). As such, the cell line may be somewhat restricted as a neuronal model. Despite this, the expression of three separate pertussis toxin-

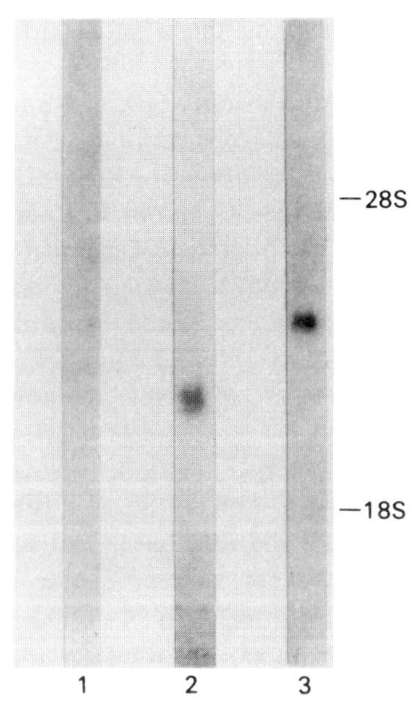

Fig. 1. *NG108-15 cells express both $G_i2\alpha$ and $G_i3\alpha$ but not $G_i1\alpha$*

Total RNA was isolated from NG108-15 cells and used in Northern blots which were probed with oligonucleotides (see [24]) corresponding to amino acids 125–135 of the α-subunits of G_i1, G_i2 and G_i3 as expressed in the rat. The $G_i2\alpha$ probe (2) identified a single band of some 2.4 kb and the $G_i3\alpha$ probe (3) a single band of 3.4 kb. The $G_i1\alpha$ probe (1) did not identify a transcript in these cells. Positions of the 18 and 28 S rRNA species are indicated for reference.

sensitive G-proteins and a range of receptors, which have been demonstrated to interact with G-proteins of this subclass, makes it well suited for experiments on the specificity of interactions of receptors with pertussis toxin-sensitive G-proteins.

As it is a homogeneous cell line, then it must be anticipated that every cell will express each of these G-proteins and that, unless some unknown physical constraints apply, each G-protein should be, at least theoretically, available to interact with each of the receptors.

Molecular Identification of 'G_i' in NG108-15 Cells as G_i2

One of the most studied receptors expressed by NG108-15 cells is an opioid receptor of the δ-subtype [31]. It is well established that in membranes prepared from these cells activation of this receptor produces inhibition of adenylate cyclase activity. As various opioid ligands both inhibit adenylate cyclase and stimulate high-affinity GTPase activity with similar potencies and efficacies [31], and because the affinity of binding of opioid agonists to this receptor is reduced in the presence of poorly hydrolysed analogues of GTP, then this effect must be transduced by a G-protein. Further, as pretreatment of these cells with pertussis toxin attenuates both opioid receptor-mediated stimulation of GTPase activity [32] and inhibition of adenylate cyclase [33], then the relevant G-protein(s) must be a substrate(s) for ADP-ribosylation catalysed by this toxin.

As noted above, the expression of only G_i2, G_i3 and G_o, and not of G_i1, as pertussis toxin substrates in these cells, limits the nature of the relevant G-protein to one of these or some combination thereof. The production of anti-peptide antisera against the extreme *C*-terminal decapeptides of each of these three polypeptides (antiserum AS7 versus *C*-terminus of $G_i2\alpha$; antiserum OC1 versus *C*-terminus of $G_o\alpha$; antiserum I3B versus *C*-terminus of $G_i3\alpha$) provided selective probes for the identification of these G-proteins. (Antiserum AS7 will identify the α-subunits of both G_i1 and G_i2 equally as these polypeptides have identical *C*-terminal decapeptides; however, as noted above, $G_i1\alpha$ is not expressed in NG108-15 cells and thus this antiserum can be used as a specific probe for $G_i2\alpha$ in these cells.) An IgG fraction was isolated from each of these antisera, as well as from normal rabbit serum, and preincubated with NG108-15 membranes. After the preincubation, basal and opioid peptide-stimulated high-affinity GTPase activities were assessed (Table 1). None of the antisera altered the basal GTPase activity of the membranes. However, the IgG fraction from antiserum AS7 completely prevented receptor stimulation of this activity. By contrast, the IgG fractions from both antiserum OC1 and I3B had no effect on opioid peptide-stimulation of high-affinity GTPase activity [28, 29].

To confirm further a direct interaction of the δ-opioid receptor with G_i2 in these cells, we performed a series of ligand-binding experiments with ^{3}H-labelled [D-Ala^2, Leu^5]enkephalin ([^{3}H]DADLE), which is a full agonist at this receptor. In the presence of Mg^{2+} (20 mM), the specific binding of [^{3}H]DADLE (approx. 400 fmol/mg of protein; K_d 2 nM) appeared to be to a single class of sites. The inclusion of the poorly hydrolysed analogue of GTP, guanosine 5′-[$\beta\gamma$-imido]triphosphate (Gpp[NH]p), (100 μM) reduced the affinity of specific [^{3}H]DADLE binding to the membranes, but did not alter the total number of sites. Similarly, pretreatment of the cells with pertussis toxin

Table 1. *Antibodies against the C-terminal decapeptides of the pertussis toxin-sensitive G-proteins expressed in NG108-15 cells: effects on basal and opioid peptide-stimulated GTPase activity*

IgG fractions were produced from each of normal rabbit serum and antisera AS7, OC1 and I3B by chromatography on protein A. Dilutions of these preparations (1 : 100) were incubated with membranes of NG108-15 cells for 1 h and then both basal high-affinity GTPase and the DADLE (10 μM) stimulation over and above the basal GTPase rate were assessed. Data are adapted from [28,29] and indicate the coupling of the opioid receptor to G_i2 in these cells.

	High-affinity GTPase activity (pmol/min per mg of protein)	
Antiserum	Basal	Opioid stimulation above basal
Normal rabbit serum	9.9 ± 0.2	4.7 ± 0.3
AS7	10.5 ± 0.2	0.1 ± 0.1
OC1	10.5 ± 0.2	4.1 ± 0.3
I3B	10.4 ± 0.2	4.6 ± 0.4

Table 2. *Antibodies against the C-terminal decapeptides of the pertussis toxin-sensitive G-proteins expressed in NG108-15 cells: effects on the specific binding of [^{3}H]DADLE*

Membranes of NG108-15 cells were incubated with IgG fractions generated as in Table 1 from normal rabbit serum and from antisera AS7, OC1 and I3B for 1 h. The specific binding of a single concentration of [^{3}H]DADLE (2 nM) which is close to the apparent K_d for this ligand was then assessed. Data are modified from [28,29]. These results demonstrate that the opioid receptor interacts specifically with G_i2 in these cells.

Antiserum	[^{3}H]DADLE specifically bound (fmol/mg of protein)
Normal rabbit serum	183.0 ± 4.4
AS7	112.5 ± 5.8
OC1	176.0 ± 9.9
I3B	173.0 ± 9.5

led to a reduction in the affinity of binding of the opioid ligand. Pretreatment of NG108-15 cell membranes with the IgG fractions of antisera AS7, OC1 and I3B produced results consistent with those of the GTPase assay (Table 2). The affinity of [^{3}H]DADLE for the receptor was reduced by the IgG fraction of antiserum AS7, but not by the equivalent fractions of antisera OC1 or I3B [28, 29]. The reduction in binding of a single, subsaturating, concentration of [^{3}H]DADLE produced by the IgG fraction of AS7 was the same as that produced by either Gpp[NH]p or by pretreatment of the cells with pertussis toxin, both of which treatments prevent interaction of receptors and pertussis toxin-sensitive G-proteins, as described above. Moreover, the IgG fraction of antiserum AS7 was unable to reduce further the specific binding of [^{3}H]DADLE produced by either pertussis toxin or Gpp[NH]p. As each of the treatments converted the entire opioid receptor population to a form which displayed reduced affinity for the ligand, then it was concluded that the pertussis toxin-sensitive G-protein linked to the δ-opioid receptor was G_i2. Uncoupling of the opioid receptor from the G-protein signalling system with the IgG fraction of antiserum AS7 also prevented opioid peptide-mediated inhibition of adenylate cyclase, further defining the true 'G_i' of the adenylate cyclase cascade in NG108-15 cells to be the product of the G_i2 gene [28,29].

Identification of the G-Protein which Transduces Receptor Regulation of Ca^{2+} Channels as G_o

After cyclic AMP-induced 'differentiation' of NG108-15 cells, it is possible to measure an α_2-adrenergic receptor-mediated depression of voltage-dependent Ca^{2+} current [34]. This response is abolished by pretreatment of the cells with pertussis toxin, implicating the involvement of a pertussis toxin-sensitive G-protein in transducing this signal. To assess the molecular identity of the relevant G-protein, we have used the selective anti-G-protein antisera defined above. Noradrenaline, at maximally effective concentrations (5 μM), was able to depress the Ca^{2+} current by some 26%. Injection of antiserum AS7 into such cells did not significantly affect the action of adrenaline, neither did intracellular injection of antiserum I3B, but intracellular injection of antiserum OC1 reduced substantially α_2-adrenergic inhibition of the Ca^{2+} current (Table 3). These results provide strong evidence for a specific role for G_o in regulating Ca^{2+} channel function [27]. Similar conclusions had been reached by Hescheler *et al.* [35], who reconstituted the α-subunits of either G_o or 'G_i' (as the G-proteins were isolated from brain then the 'G_i' was probably largely G_i1) into pertussis toxin-pretreated NG108-15 cells which had been differentiated by treatment with dibutyryl cyclic AMP. They were able to demonstrate a reconstitution of opioid peptide-mediated depression of Ca^{2+} current and that the G_o fraction was some 10 times as effective as 'G_i'.

The apparent ability of the δ-opioid receptor to interact either specifically or at least with a high degree of selectivity with G_i2 in undifferentiated NG108-15 cells, but to interact with G_o as well as G_i2 in 'differentiated' NG108-15 cells requires discussion. 'Differentiation' of NG108-15 cells is necessary to be able to record receptor regulation of Ca^{2+} currents and we have noted that associated with the morphological differentiation (usually produced by prolonged exposure of the cells to a range of pharmacological agents which are able to elevate intracellular cyclic AMP concentrations), levels of membrane-associated $G_o\alpha$ are elevated significantly [36]. Furthermore, $G_o\alpha$ does not appear to be represented by a single entity in these cells, as two distinct forms can be resolved by either two-dimensional electrophoresis or in one-dimensional SDS/PAGE when 4 M deionized urea is included in the gels. We have been unable to discriminate between these two forms using three distinct anti-peptide antisera directed against various regions of the primary structure of $G_o\alpha$. As such, these two forms may be equivalent to G_o and G_o^* identified initially by Goldsmith *et al.* [37] in bovine brain. The increase in levels of $G_o\alpha$ upon differentiation of NG108-15 cells appears to be largely restricted to the more acidic form and hence is likely to be G_o rather than G_o^* [38]. We have, however, been unable to provide any data which would indicate that these two forms of G_o differ in their degree of post-translational modification. Both forms are substrates for pertussis toxin-catalysed ADP-ribosylation and are myristoylated. Moreover, their electrophoretic mobility in isoelectric focusing gels is not modified by treatment with phosphatases.

The pertussis toxin-sensitive G-proteins are not traditionally considered to be substrates for cholera toxin. TD1, however, is a substrate for either per-

Table 3. *Effect of intracellular injection of antibodies from antisera AS7 and OC1 on noradrenaline-induced depression of voltage-dependent Ca^{2+} currents in chemically differentiated NG108-15 cells*

Data which are presented as means ± S.E.M. are adapted from [27]. These results indicate that receptor regulation of voltage-operated Ca^{2+} channels in these cells is mediated via G_o.

Antiserum	Depression of the peak inward Ca^{2+} current produced by noradrenaline (5 μM) (%)
Normal rabbit serum	26.0 ± 4.6
AS7	20.5 ± 2.5
OC1	9.0 ± 2.5

tussis or cholera toxins under appropriate conditions. Furthermore, the α-subunits of all of the identified G-proteins have an invariant arginine residue in a position equivalent to that which has been proposed to be the site of cholera toxin-catalysed ADP-ribosylation in $G_s\alpha$. When cholera toxin-catalysed ADP-ribosylation was performed on membranes of undifferentiated NG108-15 cells in the absence of exogenously supplied guanine nucleotides, then, as well as the 44 and 42 kDa polypeptides which correspond to the forms of $G_s\alpha$ in these cells, a polypeptide of some 40 kDa was labelled [39]. The 40 kDa polypeptide was not labelled when GTP was included in the assay. Inclusion of DADLE in the absence of guanine nucleotides markedly increased the radioactivity incorporated into this band in the presence of cholera toxin (Fig. 2), and dose–response curves indicated a similar K_a for this response as for DADLE stimulation of high-affinity GTPase activity and inhibition of adenylate cyclase [39]. Further, a series of opioid ligands of differing intrinsic activity, when used at maximally effective concentrations, promoted cholera toxin-catalysed ADP-ribosylation of the 40 kDa polypeptide(s) to a degree consistent with their ability to stimulate high-affinity GTPase activity (G. Milligan & F. R. McKenzie., unpublished work).

As pretreatment of the cells with pertussis toxin prevented subsequent DADLE stimulation of the cholera toxin-catalysed ADP-ribosylation of the 40 kDa polypeptide, then it is likely that this polypeptide is the α-subunit of the δ-opioid receptor-linked G-protein. Specific identification of this polypeptide as either $G_i2\alpha$, $G_i3\alpha$ or $G_o\alpha$ has required the development of gel conditions able to resolve adequately the radiolabelled polypeptides. This has recently

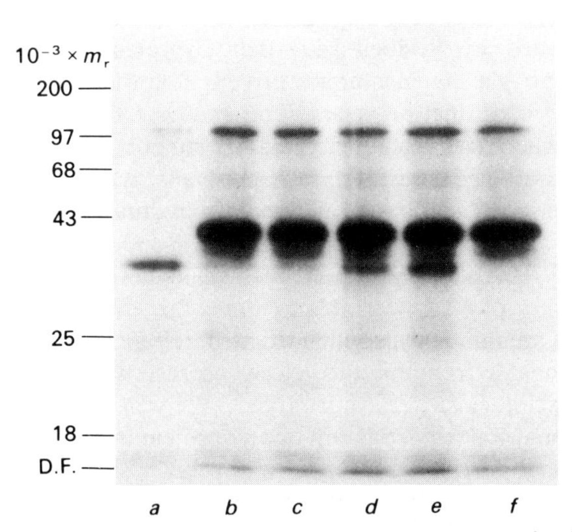

Fig. 2. *Pharmacology of the opioid receptor stimulation of cholera toxin-catalysed ADP-ribosylation of a 'G_i-like' protein in membranes of NG108-15 cells*

Membranes of NG108-15 cells were treated with either pertussis toxin (lane *a*) or cholera toxin (lanes *b–f*) in the absence of guanine nucleotides for 2 h at 37°C. Additionally, lanes contained no ligand (*a*, *f*); 1 mM-naloxone (*b*); 1 mM-naloxone + 0.1 μM-DALAMID ([D-Ala^2]leucine enkephalinamide) (*c*); 0.1 μM-DALAMID (*d*); and 10 μM-DALAMID (*e*). The Figure is taken from [39] with permission. Abbreviation: DF, dye front.

been achieved [24] and DADLE shown to allow cholera toxin-catalysed ADP-ribosylation of primarily $G_i2\alpha$ in membranes of undifferentiated NG108-15 cells (G. Milligan, I. Mullaney & F. R. McKenzie unpublished work). It was possible to also detect very weak labelling of $G_o\alpha$ in these experiments (G. Milligan, I. Mullaney & F. R. McKenzie unpublished work). It should be possible to use specific anti-$G_i2\alpha$, $G_i3\alpha$ and $G_o\alpha$ antisera to immunoprecipitate the G-protein(s) which are ADP-ribosylated by cholera toxin in the presence of opioid ligands to define further the species of G-protein with which the opioid receptor interacts.

We are currently attempting to perform similar experiments with membranes from 'differentiated' NG108-15 cells, as these express higher levels of $G_o\alpha$ than membranes derived from undifferentiated cells. However, we have previously noted that 'differentiation' of NG108-15 cells by treatment with dibutyryl cyclic AMP produces a reduction by some 50% of the number of opioid receptors [26], which may limit the use of the type of experiments outlined above.

Biochemical and Pharmacological Manipulation of Levels of G-Proteins in Cells

Alterations in the cellular complement or of the subcellular distribution of various G-proteins might be expected to have far-reaching effects on signal transduction cascades in that cell if the G-protein concentration or function was rate-limiting for the transfer of information. As such, manipulations which produce an alteration in the cellular complement of G-proteins might provide a useful approach to understanding the role(s) of the various G-proteins.

It was suggested by Rodbell [40] that G-protein α-subunits might be released from the plasma membrane of cells following receptor activation, perhaps to catalyse distinct processes at other sites within the cell. While little experimental data have been generated to support this notion, we have observed that sustained activation of G-proteins, which can be elicited in membrane fractions by incubation with poorly hydrolysed analogues of GTP, will produce a slow release of the α- but not β-subunits of the various G-proteins [41,42]. This effect was specific for analogues of GTP as neither analogues of GDP nor of ATP were able to mimic this effect. Because of the slow time courses of guanine nucleotide-mediated release, and because receptor activation was unable to mimic this release process, we do not believe that it represents a physiologically relevant process. We were, however, interested to examine whether sustained activation of a G-protein in a cell would produce a similar release and whether this might alter cellular signalling processes. The most convenient route to persistently activate a particular G-protein in a cell is to use cholera toxin. This toxin catalyses the ADP-ribosylation of $G_s\alpha$ on an arginine residue which is close to the guanine-nucleotide-binding site. Such ADP-ribosylation activates $G_s\alpha$ by inhibiting its intrinsic GTPase activity. When we treated a range of cells, including L6 skeletal myoblasts, NG108-15 neuroblastoma × glioma hybrids and C6 glioma cells with cholera toxin, immunologically detectable levels of $G_s\alpha$ in membranes from these cells was substantially reduced [43] (Fig. 3). This effect was dependent upon both the concentration of toxin used and the time of exposure. The loss of $G_s\alpha$ was not related to cholera toxin-induced increases in levels of intracellular cyclic AMP, as forskolin, 8-bromo-cyclic AMP and hormones which activate adenylate cyclase via membrane-associated receptors were ineffective [43]. Loss of $G_s\alpha$ in cholera toxin-treated cells was, however, a 'down-regulation' rather than a release as noted in membrane fractions, as the $G_s\alpha$ could not be detected in the cytoplasm of the cells. This may be a reflection of a rapid proteolytic degradation [44] subsequent to activation and release. At least in rat glioma C6 cells, in contrast to the loss of $G_s\alpha$ polypeptide, mRNA corresponding to $G_s\alpha$ was not reduced in the cholera toxin-treated cells (G. Milligan, I. C. Carr, C. Loney & J. T. Knowler unpublished work) indicating that the effect was unlikely to have been produced by transcriptional regulation. It is only $G_s\alpha$ which is reduced in amount by cholera toxin treatment. In a range of systems, levels of each of $G_i2\alpha$-, $G_i3\alpha$-, $G_o\alpha$- and β-subunit were not altered ([44,45] and G. Milligan, I. C. Carr & F. R. McKenzie unpublished work).

It might be presumed that if $G_s\alpha$ was initially limiting for signal tranduction or became so due to its 'down-regulation' after cholera toxin treatment, then the maximal degree of signalling from G_s to adenylate cyclase would be reduced in membranes of cells which had been treated chronically with cholera toxin. Indeed, in NG108-15 cells this has been demonstrated to be the case (Table 4). In measuring either basal or forskolin-amplified adenylate cyclase activity, increasing times of incubation with cholera toxin or increasing concentrations of the toxin initially produced a stimulation of adenylate cyclase activity owing to ADP-ribosylation and hence activation of $G_s\alpha$, but a sub-

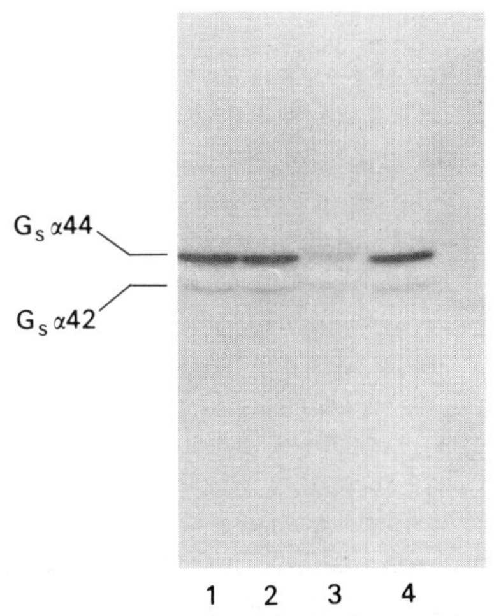

Fig. 3. *Cholera toxin treatment reduces the levels of membrane-associated $G_s\alpha$*

L6 skeletal myoblasts were treated with the pertussis toxin vehicle (lane 1); pertussis toxin (100 ng/ml) (2); cholera toxin (100 ng/ml) (3); or with the cholera toxin vehicle (4) for 24 h. Membranes from these cells (200 μg) were immunoblotted using antiserum CS1 (directed against the *C*-terminal decapeptide of $G_s\alpha$) as the primary reagent. Cholera toxin treatment markedly reduced the membrane levels of the 44 kDa form of $G_s\alpha$ particularly, but also of the 42 kDa form. The Figure is taken from [43] with permission.

Table 4. *Effects of cholera toxin on basal adenylate cyclase activity in NG108-15 cells*

NG108-15 cells were incubated with varying concentrations of cholera toxin for 24 h. The cells were then harvested, membranes were prepared and adenylate cyclase activity was assessed. Data are adapted from [45].

Cholera toxin concentration (ng/ml)	Basal adenylate cyclase activity (pmol/min per mg of protein)
0	1.7 ± 0.4
10	36.5 ± 1.0
1000	8.1 ± 1.3

sequent reduction of these elevated activities, presumably due to the loss of membrane-associated $G_s\alpha$ [45]. By contrast, in GH3 cells, Chang & Bourne [44] were unable to note a reduction in adenylate cyclase activity even with the loss of some 90% of $G_s\alpha$, which indicates that $G_s\alpha$ is apparently present in a considerable excess over the amount strictly required for maximal function in these cells. The availability of the *cyc*$^-$ variant of the murine S49 lymphoma cell line, which is genetically deficient in expression of $G_s\alpha$, allowed an examination of whether a direct effect of 'α_i' or a Mass–Action inhibition of $G_s\alpha$ activation by $\beta\gamma$-subunits, which would be generated by the dissociation of holomeric 'G_i', was the more important component in receptor-mediated inhibition of adenylate cyclase. Somatostatin was able to mediate inhibition of adenylate cyclase in membranes of *cyc*$^-$ cells [46], indicating that '$G_i\alpha$' could inhibit directly. Removal of a proportion of the $G_s\alpha$ from the membrane of a cell using cholera toxin treatment may allow an alternative approach to this question which will not be dogged by the limitation of the use of a 'mutant' cell for the assay.

Pertussis toxin treatment of a cell does not produce an equivalent down-regulation of G-proteins which are substrates for this toxin. Indeed pharmacological regulation of the levels of these G-proteins has been difficult to achieve. However, sustained stimulation of rat adipocytes with the adenosine analogue N^6-phenylisopropyl adenosine (PIA) produces an almost complete loss of the α-subunits of both G_i1 and G_i3, as well as a less marked loss of $G_i2\alpha$ [47]. By contrast, PIA treatment caused no effect on levels of either the large or small forms of $G_s\alpha$. The mechanism of adenosine A_1 receptor-activated loss of pertussis toxin-sensitive G-proteins is currently unknown, as is whether activation of A_1 receptors in other locations will produce similar results, but a series of other receptors which mediate inhibition of adenylate cyclase are expressed in rat adipocytes. As such, the functionality or otherwise of such receptors may provide insights into the roles of the different pertussis toxin-sensitive G-proteins expressed in these cells.

Conclusions

The availability of a range of techniques to examine the distribution of the various G-proteins and their interactions with both detector and effector systems have allowed rapid advances in our understanding of the function and specificity of interactions of the individual G-proteins. The use of site-directed antisera, construction of various G-protein chimeras [48], site-directed mutagenesis of expressed cDNA clones [49] and methods to manipulate the cellular complement of G-proteins, suggest means to define further the distribution and function of these proteins.

References

1. Gilman, A. G. (1987) *Annu. Rev. Biochem.* **56**, 615–649
2. Milligan, G. (1989) *Curr. Opin. Cell Biol.* **1**, 196–200
3. Gao, B., Gilman, A. G. & Robishaw, J. D. (1987) *Proc. Natl. Acad. Sci. U.S.A.* **84**, 6122–6125
4. Gautam, N., Baetscher, M., Aebersold, R. & Simon, M. I. (1989) *Science* **244**, 971–974
5. Jelsema, C. L. & Axelrod, J. (1987) *Proc. Natl. Acad. Sci. U.S.A.* **84**, 3623–3627
6. Katada, T., Northup, J. K., Bokoch, G. M., Ui, M. & Gilman, A. G. (1984) *J. Biol. Chem.* **259**, 3578–3585
7. Whiteway, M., Houghan, L., Dignard, D., Thomas, D. Y., Bell, L., Saari, G. C., Grant, F. G., O'Hara, P. & MacKay, V. L. (1989) *Cell (Cambridge, Mass.)* **56**, 467–477
8. Gierschik, P., Milligan, G., Pines, M., Goldsmith, P., Codina, J., Klee, W. & Spiegel, A. (1986) *Proc. Natl. Acad. Sci. U.S.A.* **83**, 2258–2262
9. Homberger, V., Brabet, P., Audigier, Y., Pantaloni, C., Bockaert, J. & Rouot, B. (1987) *Mol. Pharmacol.* **31**, 313–319
10. Dietzel, C. & Kurjan, J. (1987) *Cell (Cambridge, Mass.)* **50**, 1001–1010
11. Provost, N. M., Somers, D. E. & Hurley, J. B. (1988) *J. Biol. Chem.* **263**, 12070–12076
12. Kumagai, A., Pupillo, M., Gundersen, R., Miake-Lye, R., Devreotes, P. N. & Firtel, R. A. (1989) *Cell (Cambridge, Mass.)* **57**, 265–275
13. Millner, P. A. & Robinson, P. S. (1989) *Cell. Signalling* **1**, 421–433
14. Schimz, A., Hinsch, K. D. & Hildebrand, E. (1989) *FEBS Lett.* **249**, 59–61
15. Wakelam, M. J. O., Davies, S. A., Houslay, M. D., McKay, I., Marshall, C. J. & Hall, A. (1986) *Nature (London)* **323**, 173–176
16. Milligan, G. & Wakelam, M. J. O. (1989) *Prog. Growth Factor Res.* **1**, 161–177,
17. Nishimoto, I., Murayama, Y., Katada, T., Ui, M. & Ogata, E. (1989) *J. Biol. Chem.* **264**, 14029–14038
18. Reference deleted
19. Milligan, G. (1988) *Biochem. J.* **255**, 1–13
20. Robishaw, J. D., Smigel, M. D. & Gilman, A. G. (1986) *J. Biol. Chem.* **261**, 9587–9590
21. Katada, T. & Ui, M. (1982) *J. Biol. Chem.* **257**, 7210–7216
22. Jones, D. T. & Reed, R. R. (1987) *J. Biol. Chem.* **262**, 14241–14249
23. Scherer, N. M., Toro, M.-J., Entman, M. L. & Birnbaumer, L. (1987) *Arch. Biochem. Biophys.* **259**, 431–440
24. Mitchell, F. M., Griffiths, S. L., Saggerson, E. D., Houslay, M. D., Knowler, J. T. & Milligan, G. (1989) *Biochem. J.* **262**, 403–408
25. Goldsmith, P., Gierschik, P., Milligan, G., Unson, C. G., Vinitsky, R., Malech, H. & Spiegel, A. (1987) *J. Biol. Chem.* **262**, 14683–14688
26. Mullaney, I., Magee, A. I., Unson, C. G. & Milligan, G. (1988) *Biochem. J.* **256**, 649–656
27. McFadzean, I., Mullaney, I., Brown, D. A. & Milligan, G. (1989) *Neuron* **3**, 177–182
28. McKenzie, F. R. & Milligan, G. (1990) *Biochem. J.* **267**, 371–378
29. Milligan, G., Mitchell, F. M., Mullaney, I., McClue, S. J. & McKenzie, F. R. (1990) in *Hormone Perception and Signal Transduction in Animals and Plants* (Roberts, J., Venis, M. & Kirk, C.), Company of Biologists, London, in the press
30. Hamprecht, B., Glaser, T., Reiser, G., Bayer, G. & Probst, F. (1985) *Methods Enzymol.* **109**, 316–341
31. Koski, G. & Klee, W. A. (1981) *Proc. Natl. Acad. Sci. U.S.A.* **78**, 4185–4189
32. Burns, D. L., Hewlett, E. L., Moss, J. & Vaughan, M. (1983) *J. Biol. Chem.* **258**, 1435–1438
33. Kurose, H., Katada, T., Amano, T. & Ui, M. (1983) *J. Biol. Chem.* **258**, 4870–4875
34. Docherty, R. J. & McFadzean, I. (1989) *Eur. J. Neurosci.* **1**, 132–140
35. Hescheler, J., Rosenthal, W., Trautwein, W. & Schultz, G. (1987) *Nature (London)* **325**, 445–447
36. Mullaney, I. & Milligan, G. (1989) *FEBS Lett.* **244**, 113–118
37. Goldsmith, P., Backlund, P. S., Rossiter, K., Carter, A., Milligan, G., Unson, C. G. & Spiegel, A. (1988) *Biochemistry* **27**, 7085–7090

38. Mullaney, I. & Milligan, G. (1990) *Biochem. Soc. Trans.* **18**, 396–399
39. Milligan, G. & McKenzie, F. R. (1988) *Biochem. J.* **252**, 369–373
40. Rodbell, M. (1985) *Trends Biochem. Sci.* **10**, 461–464
41. Milligan, G., Mullaney, I., Unson, C. G., Marshall, A. M., Spiegel, A. M. & McArdle, H. (1988) *Biochem. J.* **254**, 391–396
42. Milligan, G. & Unson, C. G. (1989) *Biochem. J.* **260**, 837–841
43. Milligan, G., Unson, C. G. & Wakelam, M. J. O. (1989) *Biochem. J.* **262**, 643–649
44. Chang, F.-H. & Bourne, H. R. (1989) *J. Biol. Chem.* **264**, 5352–5357
45. Macleod, K. G. & Milligan, G. (1990) *Cell. Signalling* **2**, 139–151
46. Jakobs, K. H., Aktories, K. & Schultz, G. (1983) *Nature* (*London*) **303**, 177–178
47. Green, A., Johnson, J. L. & Milligan, G. (1990) *J. Biol. Chem.* in the press
48. Masters, S. B., Sullivan, K. A., Miller, R. T., Beiderman, B., Lopez, N. G., Ramachandran, J. & Bourne, H. R. (1988) *Science* **241**, 448–451
49. Landis, C. A., Masters, S. B., Spada, A., Pace, A. M., Bourne, H. R. & Vallar, L. (1989) *Nature* (*London*) **340**, 692–696

Biochem. Soc. Symp. **56**, 35–44
Printed in Great Britain

The Molecular Basis of GTP-Binding Protein Interaction with Receptors

HEIDI E. HAMM, HELEN RARICK, MARIA MAZZONI,
JUSTINE MALINSKI and KYONG-HOON SUH

Department of Physiology and Biophysics, University of Illinois College of Medicine at Chicago, Box 4998, Chicago, IL 60680, U.S.A.

Synopsis

The molecular basis of the interaction between the visual receptor, rhodopsin, and the rod outer segment GTP-binding protein, transducin or G_t, was studied using a synthetic-peptide-competition approach to elucidate the site(s) on the G_t α-subunit (α_t) involved in high-affinity binding to light-activated rhodopsin (R*). Synthetic peptides based on the amino acid sequence of portions of the molecule that interact with rhodopsin can themselves bind the rhodopsin and thus behave as competitive inhibitors of rhodopsin–G-protein interaction. This blockade was assessed by measuring the ability of peptides to inhibit G_t stabilization of the metarhodopsin II conformation of rhodopsin. Based upon this analysis, two regions near the *C*-terminal of α_t are important for interaction with R*.

GTP-Binding Protein Interaction with Receptors

The role of GTP-binding proteins in signal transduction by a variety of different sensory and hormone receptors is an area of intense current investigation. The mechanisms of receptor interaction with and activation of G-proteins is the subject of this review. The visual transduction system of the vertebrate retina presents some unique advantages for molecular studies of receptor–G-protein interaction. In the photoreceptor cells, rods and cones, light activation of rhodopsin causes a hyperpolarization of the plasma membrane by activating a biochemical cascade that lowers cyclic GMP levels and causes channel closure [1]. Light-activated rhodopsin has this effect by activating G_t which in turn activates cyclic GMP phosphodiesterase (PDE). Studies of the rod response to light have shown that at least 1000 G-protein molecules can be activated by each activated rhodopsin [1], and activation of one G-protein occurs within a millisecond [2]. Within this time, the activated receptor binds to a G-protein and two conformational changes occur in the G-protein, the first to open the guanine-nucleotide-binding pocket, releasing GDP, and the second, GTP-dependent large change leading to loss of affinity for receptor and for $\beta\gamma$-subunit, and increased affinity for PDE [3]. One reason that this system is particularly suited for studying receptor–G-protein interaction, is that the interaction itself can be directly measured spectroscopically. This is because the visual receptor contains a chromophore, retinal, the light absorption spectra of which change with the changes in rhodopsin conformation during excitation. G_t binds specifically to the metarhodopsin II conformation and this binding can be detected spectroscopically under appropriate conditions [4].

The present studies originated with an interest in the structure–function relationships of the α-subunit, including sites of interaction with the $\beta\gamma$-subunit, rhodopsin and PDE. The studies were aimed at testing a model of the α-subunit presented by Deretic & Hamm [5] (Fig. 1*b*), based on secondary structure prediction, the tertiary structure of the GDP-binding pocket of EF-Tu solved by Jurnak [6] and competition studies between various proteins that recognize the α-subunit ($\beta\gamma$, rhodopsin, 4A, pertussis toxin and trypsin) [7].

Fig. 1(*a*) summarizes what is known about functional sites on the G_t α-subunit. The *N*-terminal is important for interaction with the $\beta\gamma$-subunit. Removal of the *N*-terminal region of α_t by *Staphylococcus aureus* V8 proteinase impairs $\beta\gamma_t$ binding and the presence of $\beta\gamma$ slows down proteolysis [8]. However, it is not clear whether the *N*-terminal region of the α-subunit interacts with the $\beta\gamma$-subunit directly or whether its presence maintains some conformation of the α-subunit that is required for heterotrimer formation. Another reason to wonder if the *N*-terminal is the only site of $\beta\gamma$ interaction is that one might expect a conserved sequence in various α-subunits that interact with the same $\beta\gamma$-subunits, and there is very little conservation of sequences in

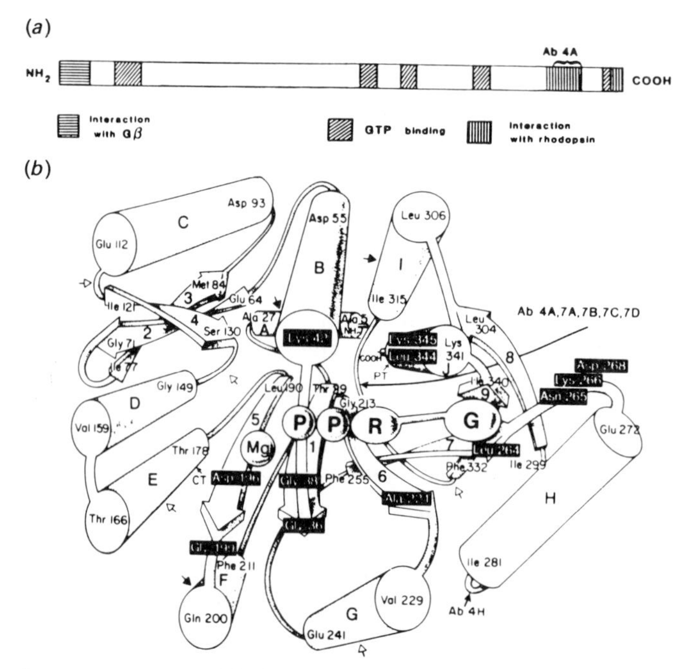

Fig. 1. (*a*) *Linear representation of the functional domains on the α-subunit of* G_t *and* (*b*) *a model of the three-dimensional structure of α_t*

From Deretic & Hamm [5] with permission.

the *N*-terminal region. There is some evidence that other parts of the α-subunit take part in the $\beta\gamma$ interaction site. Hingorani *et al.* [9], studying the chemically cross-linked products of G_t-subunits, have shown that the 5 kDa *C*-terminus fragment of α_t directly interacts with β_t. Further, Kahn & Gilman [10] have shown that cholera toxin-mediated ADP-ribosylation of Arg-196 of the α-subunit changes its affinity for $\beta\gamma$, suggesting that this region might also take part in the $\beta\gamma$-binding site.

The *C*-terminal is important for interaction with receptor. ADP-ribosylation of Cys-347 at the *C*-terminal by pertussis toxin uncouples G_t (as well as other pertussis toxin substrates) from receptor [11,12]. A series of monoclonal antibodies that block G_t interaction with rhodopsin also map to the *C*-terminal 5 kDa region of α_t [5,7]. Additional evidence for the importance of the *C*-terminal region in receptor interaction is the *unc* mutation in S49 lymphoma cells which changes Arg-390 to Pro [13]. In addition, as will be summarized in this review, synthetic peptides corresponding to two regions of the *C*-terminal of α_t bind to rhodopsin and block rhodopsin–G_t interaction.

The binding region for interaction with effectors is unknown at the present time. A reasonable speculation is that areas that change conformation from the inactive, GDP-bound, α-subunit to the GTP-bound, active, subunit are involved in effector interaction, since the GDP-bound, inactive, α-subunit does not activate effectors. (However, it is not excluded that it interacts with them.) In the two GTP-binding proteins the structures of which have been determined, EF-Tu and ras p21, at least two regions near the γ-phosphate-binding site change their conformation upon GTP binding [6,14–16, 28]. A change in conformation in the effector-binding region, residues Ser-17–Tyr-40 in p21, allows the binding of the GTPase-activating protein (GAP), the probable effector protein of p21 [17]. A second more dramatic conformational change takes place in the turn–helix–turn region just after the DXXG sequence, residues 57–60. There is an extra turn of the helix in the GTP-bound conformation, resulting in a change in conformation of the two turns at either end of the helix. These local conformational changes could of course be propagated to distant regions of the molecule. The similar regions in G_t are Gly-162–Ile-181 and Asp-196–Gly-213, respectively.

Evidence available at the present time for an effector-binding site on the α-subunits of heterotrimeric G-proteins is rather inconclusive. There is a small region of sequence similarity in the first region, the ras-effector-binding site, [18]. The significance of this sequence similarity is unknown. Functional studies with chimeric α-subunits strongly suggest that effectors interact with the *C*-terminal 40% of α-subunits [19]. More fine-tuned chimeras have confirmed this localization and placed it in a more restricted region between Phe-211 and Asp-311 [20].

A series of regions of sequence similarity with other GTP-binding proteins with a similar order of appearance as in ras p21 and EF-Tu defines the GTP-binding pocket (Fig. 1*a*). Cholera toxin-catalysed ADP-ribosylation of Arg-174, near the Mg^{2+}- and γ-phosphate-binding sites, inhibits GTPase activity [3]. Other structural features of the α_t-subunit include the tryptic cleavage sites at Lys-17, Arg-204 and Arg-310.

The approach taken in the present work was to generate monoclonal antibodies to various regions of α_t and assess the functional effect of antibody binding. A library of monoclonal antibodies (mAb) against the α_t-subunit [21] was screened for functional effects, and one particular mAb, called 4A, was shown to block light activation of PDE [22]. Studies of the mechanism of antibody blockade showed that it had no effect on α_t–PDE interaction (Fig. 2*a*), but it blocked interaction of G_t with light-activated rhodopsin, preventing G_t activation (Fig. 2*b*) [7]. It also blocked interaction of G_t with rod outer segment (ROS) membranes in the dark [7]. This could be due to blockade of G_t interaction with rhodopsin or blockade of interaction with the $\beta\gamma$-subunit would also lead to loss of interaction with rhodopsin. Epitope-mapping studies [5] and peptide-competition studies [23] determined that the main antibody-binding region is located between Asp-311 and Phe-350 at the *C*-terminal region of α_t (Fig. 3). Data outlined below show that this region is important for interaction with rhodopsin. However, we have recently shown that antibody binding also causes dissociation between G_t sub-

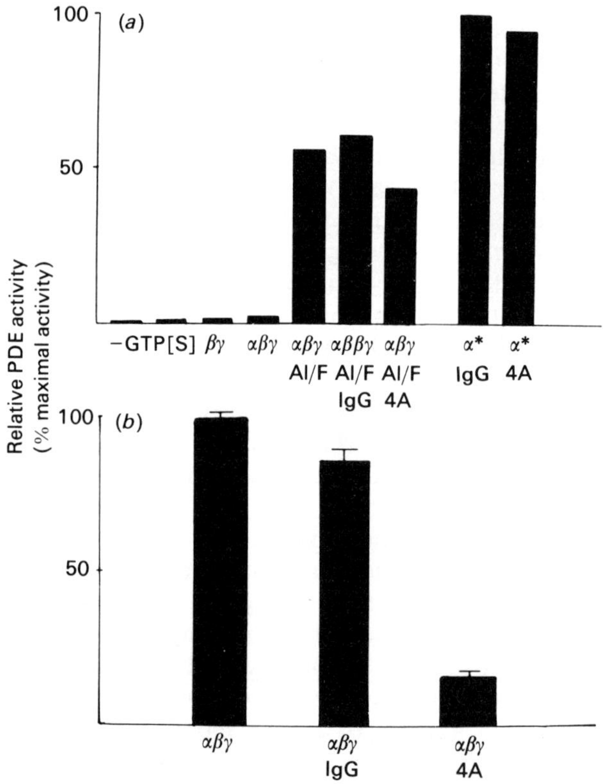

Fig. 2. *Functional effects of mAb 4A in a purified reconstituted system*

(*a*) mAb 4A has no effect on α_t–PDE interaction in an assay system consisting of cyclic GMP, purified PDE and either purified heterotrimeric G_t or purified active α_t–GTP[S](α*). (*b*) MAb 4A blocks light-activation of PDE in an assay system consisting of purified rhodopsin reconstituted in phospholipid vesicles, G_t and PDE. PDE was measured in the presence of cyclic GMP and GTP as previously described [7]. Abbreviations used: GTP[S], guanosine 5′-[γ-thio]triphosphate; AL/F, aluminium fluoride.

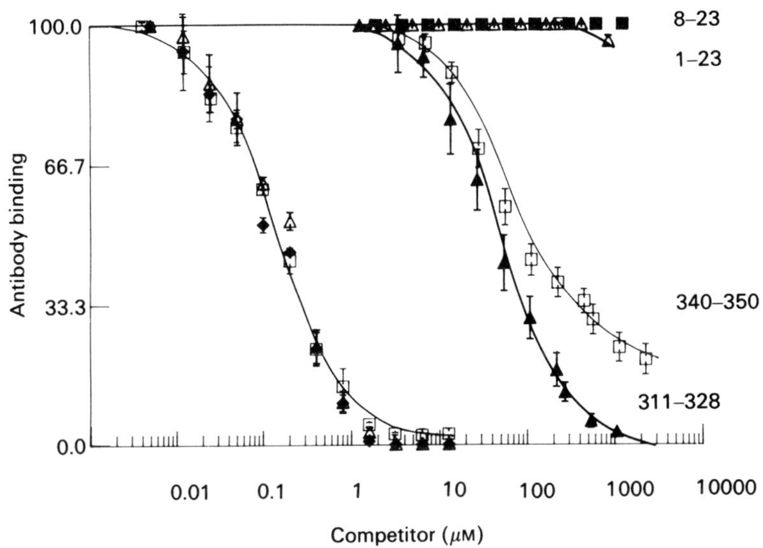

Fig. 3. *Competition between various conformational states of* α_t *and native* G_t *for binding to mAb 4A, and between native* G_t *and synthetic peptides corresponding to various regions of* α_t.
ELISA assay was performed as previously described [23]. □, G_t; ◆, α–GDP; △, α–GTP[S].

units [24]. Other antibodies that bind the *C*-terminal region of G_t also cause α–$\beta\gamma$ dissociation [24]. This appears to be a steric or conformational effect, since the antibody binds equally well to pure α-subunit, either in GDP- or GTP-bound conformations, and to the heterotrimer $\alpha(\beta\gamma)$ (Fig. 3).

If it is true that the antibody blocks interaction with rhodopsin, synthetic peptides corresponding to the antibody epitope might be expected to bind to rhodopsin. This should block G_t interaction with rhodopsin. Fig. 4 shows the effect of synthetic peptide α_t-Asp-311–Val-328 on antibody binding and interaction with rhodopsin. This peptide, and smaller fragments, block antibody binding in a competition enzyme-linked immunosorbent assay (ELISA) (Fig. 4*a*). The interaction with rhodopsin was measured using an assay of the G_t-induced stabilization of metarhodopsin II [23]. The peptide blocked rhodopsin–G_t interaction dose dependently (Fig. 4*b*). The effects of smaller fragments of this peptide are shown in the lower panels.

The role of two other regions of α_t, the *C*- and *N*-termini, were also examined in these two assays. The finding that synthetic peptides corresponding to the *N*-terminal have no effect on mAb 4A binding suggests that the *N*-terminus is not a part of the 4A epitope (Fig. 5*a*). *N*-Terminal peptides of different lengths do, however, have some effect on rhodopsin binding. This would suggest that *N*-terminal amino acids take part in rhodopsin binding (however, see below).

The *C*-terminal synthetic peptide α_t-Ile-340–Phe-350 blocks mAb 4A binding, demonstrating that the *C*-terminal 11 amino acids do take part in the mAb 4A epitope (Fig. 5*b*). These two parts of G_α, 311–328 and 340–350, must be close to each other in the three-dimensional structure, since both regions are part of the MAb 4A-binding site. Peptide α_t-340–350 has a biphasic effect

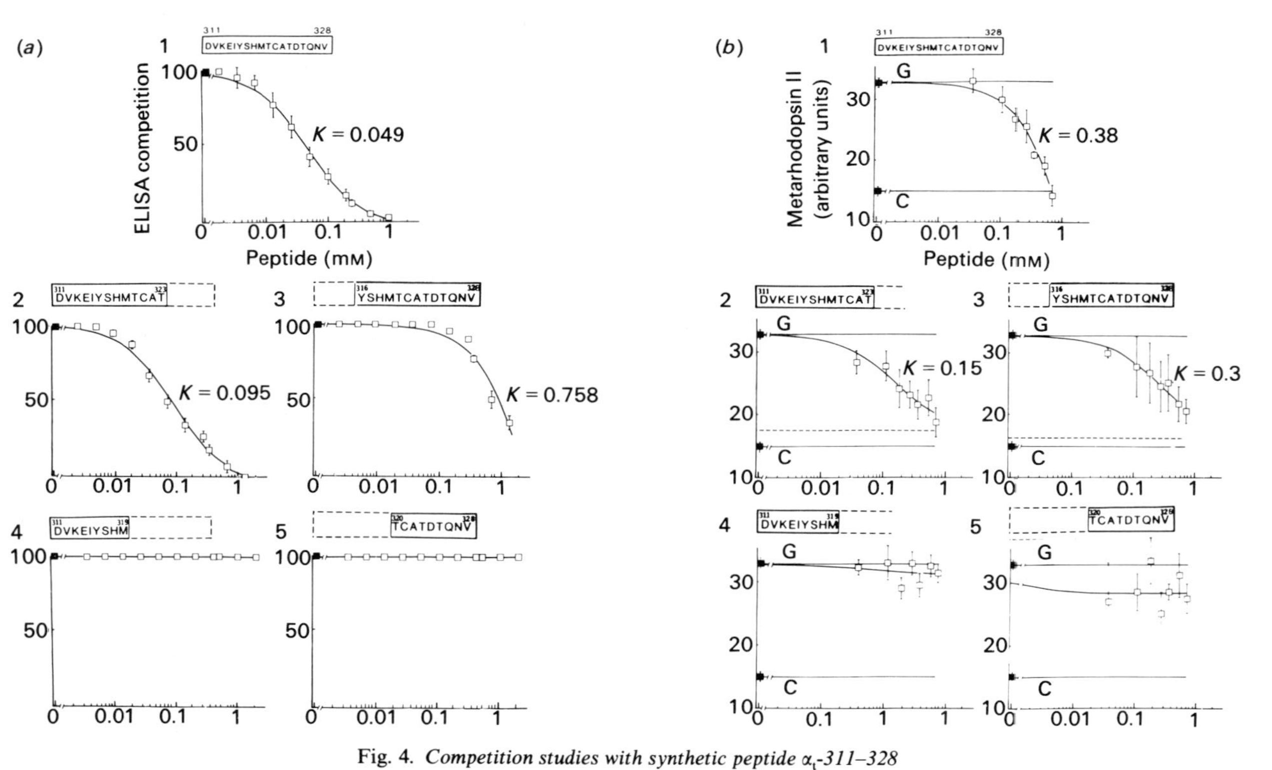

Fig. 4. *Competition studies with synthetic peptide α_t-311–328*

(*a*) Effect of synthetic peptide 311–328, corresponding to the major mAb 4A epitope, and smaller fragments on antibody–G_t interaction measured in a competition ELISA, and (*b*) effect of synthetic peptide 311–328 and smaller fragments on rhodopsin–G_t interaction as measured spectroscopically by the stabilizing effect of G_t on metarhodopsin II as previously described [23] G, in the presence of G_t; C, control metarhodopsin II in the absence of added G_t. The five panels represent data obtained with synthetic peptides of different lengths.

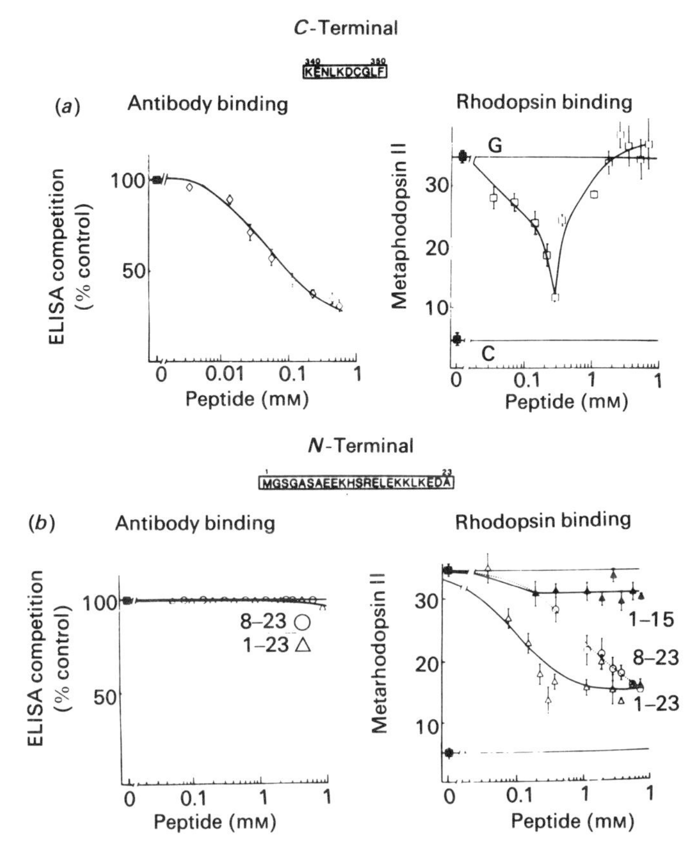

Fig. 5. *Effects of C- and N-terminal synthetic peptides on antibody–G_t and rhodopsin–G_t interaction*

(*a*) Left: effect of *C*-terminal synthetic peptides on antibody–G_t interaction measured in a competition ELISA. Right: effect of *C*-terminal synthetic peptides on rhodopsin–G_t interaction measured spectroscopically. (*b*) Left: effect of *N*-terminal synthetic peptides on antibody–G_t interaction measured in a competition ELISA. Right: effect of *N*-terminal synthetic peptides on rhodopsin–G_t interaction measured spectroscopically. G, in the presence of G_t; C, control metarhodopsin II in the absence of G_t. Metarhodopsin II is expressed in arbitrary units.

on metarhodopsin II levels, consistent with a complex effect of this peptide on metarhodopsin II. At low concentrations, it inhibits rhodopsin–G-protein interaction, while at higher concentration it causes increased metarhodopsin II levels. This peptide is one of two (of some 20 tested) that have biphasic effects. The other peptide is a modified 311–328 peptide, with its *N*-terminal acetylated and its *C*-terminal amidated (Fig. 6*a*). Compared with the parent 311–328 peptide, it is 10-fold more potent at blocking extra metarhodopsin II, but at higher concentrations it also causes increased metarhodopsin II.

It appears that these peptides may directly be able to stabilize metarhodopsin II on their own, mimicking G_t. To test this idea, the peptides were added to washed membranes in the absence of G_t. Both of these peptides also have a direct effect on rhodopsin in the absence of G_t (Fig. 6*b*). These two peptides are both more potent than the other peptides tested at blocking R*–G_t interaction, and both bind to metarhodopsin II specifically with high enough affinity at higher concentrations to stabilize metarhodopsin II, similar to G_t

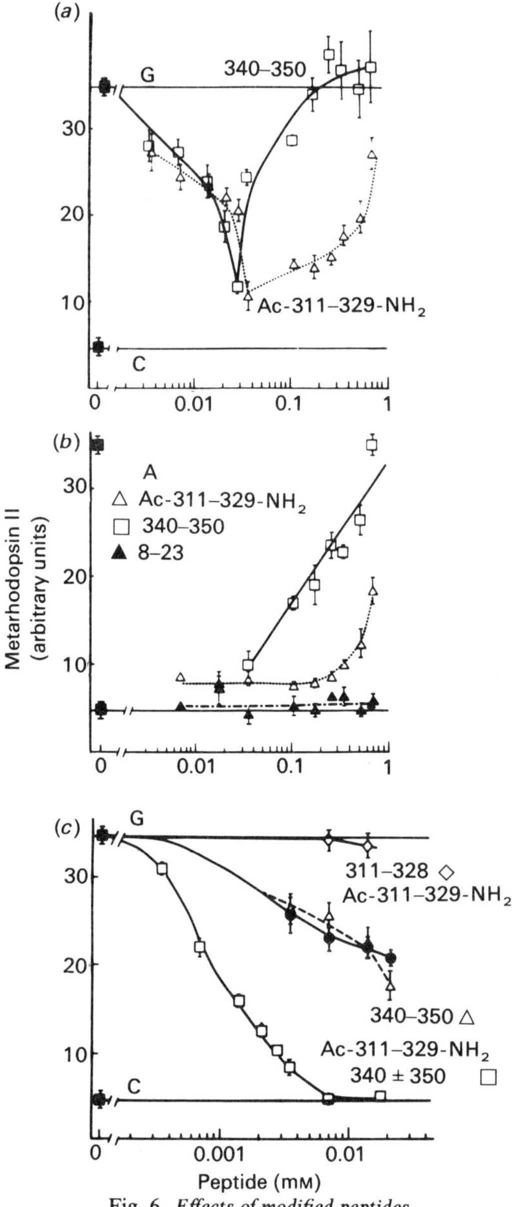

Fig. 6. *Effects of modified peptides*

(*a*) Effect of modified peptides Ac-311–329-NH_2 and Ac-305–329-NH_2 on rhodopsin–G_t interaction. (*b*) Direct effect of various synthetic peptides on metarhodopsin II. (*c*) Co-operative effects of two synthetic peptides to block rhodopsin–G_t interaction. G, in the presence of G_t; C, control metarhodopsin II in the absence of G_t.

binding. This is strong evidence that these two regions of the α-subunit directly interact with rhodopsin.

The *N*-terminal peptide α_t-Asp-8–Ala-23 does not have this effect (Fig. 6*b*); thus, this result does not support that notion that the *N*-terminal region of the α-subunit directly interacts with rhodopsin. It is possible that *N*-terminal pep-

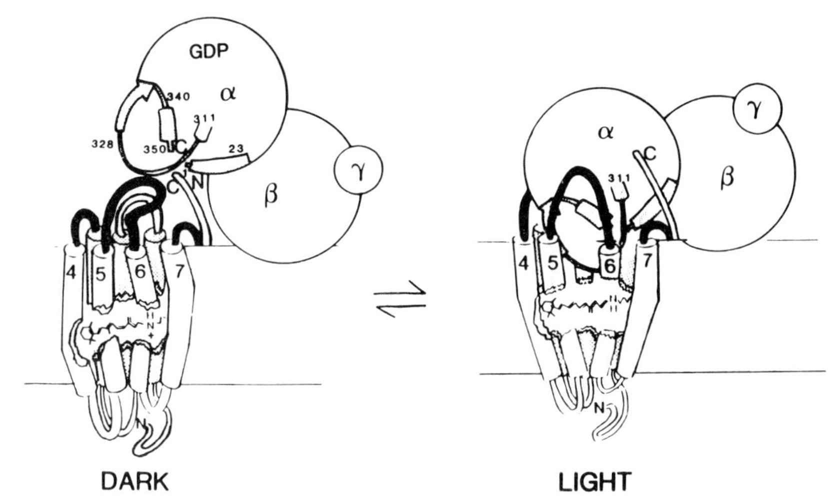

Fig. 7. *Model of the structural basis of rhodopsin–G_t interaction*

tides indirectly affect rhodopsin–G_t interaction by blocking α–$\beta\gamma$ interaction. However, one cannot rule out that the *N*-terminus is a part of the rhodopsin-binding site, as suggested by Hingorani *et al.* [9]. The fact that G_α-Ac-311–329-NH_2 and G_α-340–350 both cause direct effects on rhodopsin excludes the possibility that they inhibit G-protein binding to rhodopsin by inhibiting the interaction of the G_α- and $\beta\gamma$-subunits.

Interestingly, when the two peptides that are the most potent at competing with G_t are added together to the assay, they are 20-fold more potent than either peptide alone (Fig. 6*c*). Thus, there is a marked potentiation of the peptides in combination.

The above data argue strongly that the *C*-terminal region of α_t is directly involved in binding to rhodopsin. In support of this, the 48 kDa protein arrestin, another protein that interacts with rhodopsin, shares sequence similarity with the *C*-terminal 40 amino acids of α_t [23,25], and thus it is likely that the two proteins compete for rhodopsin.

All of this, including the fact that at least two regions on α_t bind rhodopsin, and that they have synergistic effects upon binding, suggests complex interaction with rhodopsin, including at least two linear sequences juxtaposed in the tertiary structure of the protein to form a binding site with rhodopsin.

The complementary face of this interaction, the cytoplasmic surface of rhodopsin, has been studied to elucidate the G_t-binding region using a variety of approaches, including biophysical techniques, chemical modification, proteolysis, site-directed mutagenesis of various regions on the cytoplasmic surface of rhodopsin, and other techniques (M. Applebury, unpublished work; and see reviews in [26,27]). Peptide-competition studies [29] presented evidence that three regions of rhodopsin's surface take part in binding G_t. The second and third cytoplasmic loops, and a putative fourth loop between the cytoplasmic extension of the seventh α-helix and the palmitoylated cysteines Cys-322 and -323, compete with rhodopsin for $G_t$binding [29].

The binding affinity between activated metarhodopsin II and G_t is three orders of magnitude higher than between dark-adapted rhodopsin and G_t [7,30,31]. It appears that at least three regions on rhodopsin and two on G_t make up the high-affinity binding interaction. Another region on G_t is suspected [32]. Using the above information on the regions of rhodopsin and G_t that interact, it is possible to present a model of the structural basis of rhodopsin–G_t interaction (Fig. 7). This diagram makes it clear that, in addition to electrostatic interactions between rhodopsin's cytoplasmic loops and the surface-exposed regions of G_t, hydrophobic interaction between the α-helices of rhodopsin and hydrophobic amino acids in α_t is likely to be another stabilizing force in the high-affinity rhodopsin–G-protein interaction. In fact, a hydrophobic stretch of amino acids connecting the two regions 311–329 and 340–350, α_t-Phe-332–Ile-340, is a candidate for such interaction.

References

1. Liebman, P. A., Parker, K. R. & Dratz, E. A. (1987) *Annu. Rev. Physiol.* **49**, 765–791
2. Vuong, T. M., Chabre, M. & Stryer, L. (1984) *Nature (London)* **311**, 659–661
3. Gilman, A. G. (1987) *Annu. Rev. Biochem.* **56**, 615–649
4. Emeis, D., Kuhn, H., Reichert, J. & Hofmann, K. P. (1982) *FEBS Lett.* **143**, 29–34
5. Deretic, D. & Hamm, H. E. (1987) *J. Biol. Chem.* **262**, 10831–10847
6. Jurnak, F. (1985) *Science* **230**, 32–36
7. Hamm, H. E., Deretic, D., Hofmann, K. P., Schleicher, A. & Kohl, B. (1987) *J. Biol. Chem.*, **262**, 10831–10838
8. Navon, S. E. & Fung, B. K.-K. (1987) *J. Biol. Chem.*, **262**, 15746–15751
9. Hingorani, V. N., Tobias, D. T., Henderson, J. T. & Ho, Y.-K. (1988) *J. Biol. Chem.* **263**, 6916–6926
10. Kahn, R. A. & Gilman, A. G. (1984) *J. Biol. Chem.* **259**, 6235–6240
11. Kurose, H., Katada, T., Amano, T. & Ui, M. (1983) *J. Biol. Chem.* **258**, 4870–4875
12. Cote, T. E., Frey, E. A. & Sekura, R. D. (1984) *J. Biol. Chem.* **259**, 8693–8698
13. Sullivan, K. A., Miller, R. T., Masters, S. B., Beiderman, B., Heideman, W. & Bourne, H. R. (1987) *Nature (London)* **3**[illegible], 758–762
14. de Vos, A. M., Tong, L., Milburn, M. V., Matias, P. M., Jancarik, J., Noguchi, S., Nishimura, S., Miura, K., Ohtsuka, E. & Kim, S.-H. (1988) *Science* **239**, 888–893
15. Pai, E. F., Kabsch, W., Krengel, U., Holmes, K. C., John, J. & Wittinghofer, A. (1989) *Nature (London)* **341**, 209–214
16. Jurnak, F., Heffron, S. & Bergmann, E. (1990) *Cell (Cambridge, Mass.)* **60**, 525–528
17. Schaber, M. D., Garsky, V. M., Boylan, D., Hill, W. S., Scolnick, E. M., Marshall, M. S., Sigal, I. S. & Gibbs, J. B. (1989) *Proteins: Struct. Funct. Genet.* **6**, 306–315
18. McCormick, F. (1989) *Nature (London)* **340**, 678–679
19. Masters, S. B., Sullivan, K. A., Beiderman, B., Lopez, N. G., Ramachandran, J. & Bourne, H. R. (1988) *Science* **241**, 448–451
20. Gupta, S. K., Diez, E., Heasley, L. E., Osawa, S. & Johnson, G. L. (1990) *Science* in the press
21. Witt, P. L., Hamm, H. E. & Bownds, M. D. (1984) *J. Gen. Physiol.* **84**, 251–263
22. Hamm, H. E. & Bownds, M. D. (1984) *J. Gen. Physiol.* **84**, 265–280
23. Hamm, H. E., Deretic, D., Arendt, A., Hargrave, P. A., Koening, B. & Hofmann, K. P. (1988) *Science* **241**, 832–835
24. Mazzoni, M. R. & Hamm, H. E. (1989) *Biochemistry* **28**, 9873–9880
25. Wistow, G. J., Katial, A., Craft, C. & Shinohara, T. (1986) *FEBS Lett.* **196**, 23–27
26. Findlay, J. B. C. & Pappin, D. J. C. (1986) *Biochem. J.* **238**, 625–642
27. Hofmann, K. P. (1986) *Photobiochem. Photobiophys.* **13**, 309–327
28. Milburn, M. V., Tong, L., de Vos, A. M., Brüngen, A., Yamaizumi, Z., Nishimura, S. & Kim, S. H. (1990) *Science* **247**, 939–945
29. Koenig, B., Arendt, A., McDowell, J. H., Kahlert, M., Hargrave, P. A. & Hofmann, K. P. (1989) *Proc. Natl. Acad. Sci. U.S.A.* **86**, 6878–6882
30. Bennett, N. & Dupont, Y. (1985) *J. Biol. Chem.* **260**, 4156
31. Schleicher, A. & Hofmann, K.-P. (1987) *J. Membr. Biol.* **95**, 271–281
32. Hamm, H. E. (1990) in *Advances in Second Messenger and Phosphoprotein Research* (Robison, A. & Greengard, P., eds.), vol. 24, pp. 76–81, Raven Press, New York

Biochem. Soc. Symp. **56**, 45–60
Printed in Great Britain

Direct and Indirect Modulation of Neuronal Calcium Currents by G-Protein Activation

ANNETTE C. DOLPHIN, ELAINE HUSTON and RODERICK H. SCOTT*

Department of Pharmacology, St George's Hospital Medical School, Cranmer Terrace, London SW17 0RE, U.K.

Synopsis

Evidence is presented for modulation of the various components of neuronal Ca^{2+} currents by G-proteins and by γ-aminobutyric $acid_B$ ($GABA_B$) receptor activation. The mechanism of these interactions and the possible involvement of second messengers is discussed. The role of inhibition of Ca^{2+} channels in presynaptic inhibition of neurotransmitter release is examined.

Introduction

G-proteins can modulate ion channel activity in two ways. The opening of an ion channel can be modified either directly by its association with an activated G-protein, or indirectly via a second messenger formed by the interaction of an activated G-protein with an enzyme. Both these mechanisms will be slower in onset and offset than ligand-gated processes such as acetylcholine activation of the nicotinic receptor–channel complex, which cease when the neurotransmitter is lost from the receptor. The termination of G-protein-activated processes requires GTP hydrolysis and re-association of the subunits (for reviews see Gilman, 1987; Dolphin, 1987, 1990*a*). Second messenger-mediated processes also require metabolism of the second messenger and possibly dephosphorylation if the second messenger has activated a protein kinase. Thus they are highly dependent on intracellular energy levels and temperature. One of the classes of ion channel whose activity is modified by such indirect mechanisms is the voltage-activated Ca^{2+} channel.

Voltage-activated Ca^{2+} channels are present throughout nerve cells (Llinas & Yarom, 1981; Tsien *et al.*, 1988). They are involved in many of the activities of the neuron, including the initial response to neurotransmitters. It is only rarely that neurotransmitters cause Ca^{2+} channels to open in the absence of a concomitant depolarization; however, the response may involve Ca^{2+} channels indirectly, since they may be activated after depolarization of the membrane by other transmitters such as glutamate (Choi, 1988). The activation of Ca^{2+} channels may lead to the propagation of Ca^{2+} spikes which are involved in dendritic integration (Llinas & Yarom, 1981). They also play an

* Present address: Department of Physiology, St George's Hospital Medical School, Cranmer Terrace, London SW17 0RE, U.K.

essential role in the release of neurotransmitter, which depends critically on an influx of Ca^{2+} through voltage-gated channels at the presynaptic terminal (Augustine *et al.*, 1987).

Types of Ca^{2+} Channel in Neurons

The existence of the dihydropyridine-sensitive high-voltage-activated L channel in several neuronal cell classes has now been well documented (Carbone & Lux, 1984; Nowycky *et al.*, 1985). The presence of a smaller conductance, low threshold or T channel was subsequently shown in peripheral neurons (Carbone & Lux, 1984). T channels have been reported to be blocked by octanol (Llinas, 1988), amiloride and phenytoin (for review see Tsien *et al.*, 1988), although these compounds all have other effects in the nervous system. T channels are also blocked in thalamic neurons by several other anticonvulsants (Coulter *et al.*, 1989). It is clear that, because of its threshold for activation near the resting potential, this channel contributes to membrane potential oscillations in some cells. These result from a dynamic balance between inward current through low-threshold Ca^{2+} channels and outward current through Ca^{2+}-activated channels. These oscillations will give rise to a variety of patterns of repetitive firing (Llinas & Yarom, 1981; Burlhis & Aghajanian, 1987; Coulter *et al.*, 1989).

Another distinct subtype of channel appears to be present in neurons. It was originally described in dorsal root ganglion neurons (DRGs) and has been termed an N channel. It is not dihydropyridine sensitive, and has a single channel conductance and kinetics of activation and inactivation intermediate between T and L channels (Nowycky *et al.*, 1985; Plummer *et al.*, 1989). The relative contribution of N and L channels to the high-threshold whole-cell currents in neurons is not clear cut in terms of the net resultant current being the sum of two distinct currents with differing pharmacological and biophysical properties (Swandulla & Armstrong, 1988). The problem arises because of their similar threshold for activation and the lack of a specific antagonist for N channels. In addition, although the high threshold current due to activation of N channels is thought to inactivate quite rapidly in DRGs, in sympathetic neurons N channels are less easily distinguished from L channels by their rate of voltage-dependent inactivation (Tsien *et al.*, 1988).

Modulation of High-Threshold Ca^{2+} Channels by G-Protein Activation

Many investigators have studied the interaction between Ca^{2+} currents and GTP-binding proteins. In most cases, the unequivocal demonstration of a direct interaction has not been achieved, but in the following section the possibility of direct coupling of Ca^{2+} channels to the different classes of G-proteins will be assessed.

In many cell types such as DRGs which exhibit Ca^{2+} action potentials, or in which there is a substantial Ca^{2+}-dependent plateau phase of the action potential, neurotransmitters and neuromodulators have been observed to decrease this component (Dunlap & Fischbach, 1981). An example is shown in

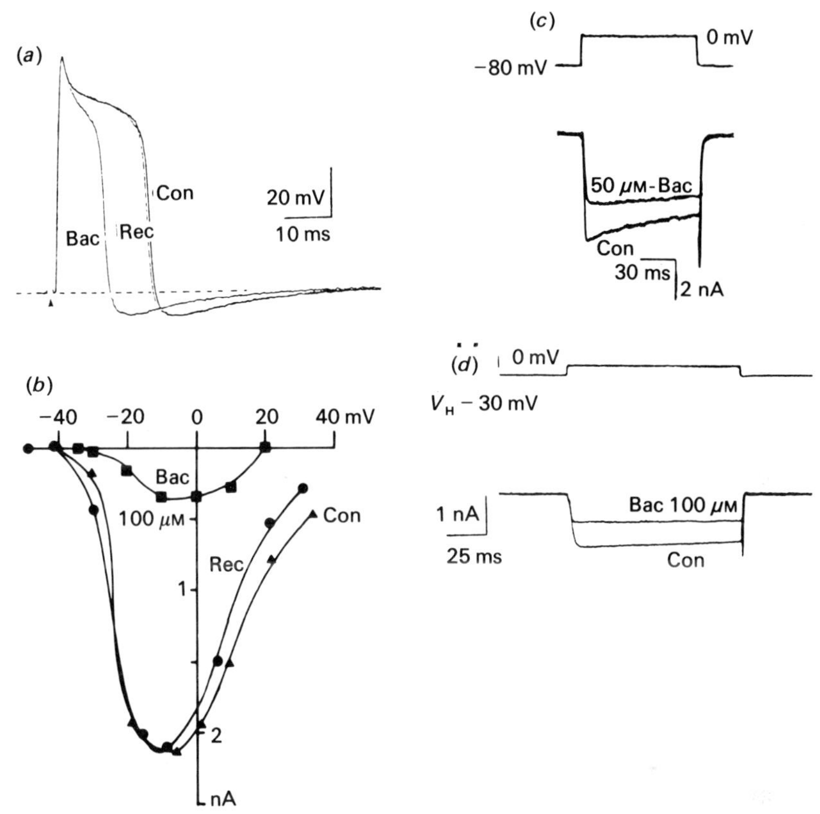

Fig. 1. *Effect of (−)-baclofen on DRG action potentials and Ca^{2+}-channel currents*

(*a*) Reduction by 100 μM-(−)-baclofen (Bac) of the action potential duration, with no effect on resting membrane potential. The action potential was evoked by a brief depolarizing pulse (arrow head) and was prolonged by 2.5 mM-tetraethylammonium. (*b*) Current–voltage relationship, showing inhibition by 100 μM-baclofen (■) of control (Con) Ca^{2+}-channel current (▲), with complete recovery (Rec) at 5 min (●). The current was carried by Ba^{2+} (2.5 mM). All currents are leak subtracted. Holding potential (V_H) − 80 mV. (*c*) Differential inhibition and slowing of the rate of activation of the transient I_{Ba} by 50 μM-baclofen. (*d*) Inhibition by 100 μM-baclofen of the sustained current available from −30 mV.

Fig. 1(*a*) of the reduction of DRG action potential duration by the $GABA_B$ agonist (−)-baclofen. Subsequently, it has been observed that Ca^{2+} currents in many cell types are inhibited by a variety of hormones and neurotransmitters. In Fig. 1(*b*), a current–voltage relationship is shown for baclofen (100 μM) inhibition of Ca^{2+} channel currents in cultured rat DRGs.

The initial evidence that a GTP-binding protein is involved was suggested by the finding that the inhibition of Ca^{2+} currents by GABA, baclofen and noradrenaline in chick and rat DRG neurons could be enhanced by GTP analogues (Scott & Dolphin, 1986), and inhibited by GDP analogues (Holz *et al.*, 1986) and by pertussis toxin. There are now many other examples in other types of neurons and neuronal cell lines (for review see Tsien *et al.*, 1988). In many cases, neurotransmitter modulation of Ca^{2+}-channel currents involves selective inhibition of the transient component of the high-threshold current.

An example is shown in Fig. 1(*c*), in which 50 μM-baclofen differentially inhibits the transient component of the high-threshold Ca^{2+}-channel current recorded in cultured rat DRGs. This has been suggested to represent inhibition of N current (for review see Tsien *et al.*, 1988). In some cases this interpretation must be regarded with caution, particularly where Ca^{2+} is the charge carrier, because of the additional complication of Ca^{2+}-dependent inactivation (Eckert & Tillotson, 1981). In some studies, it is clear that neurotransmitters have a complex effect on several components of the current; for example, a diversity of responses to noradrenaline has been observed in NG 108-15 cells (Docherty & McFadzean, 1989). Indeed, in DRG cells held at a depolarized holding potential of −30 mV, baclofen was still able to inhibit the remaining current (Fig. 1*d*), although only non-inactivating current was present, all the transient high-threshold current being inactivated at this holding potential (Dolphin & Scott, 1990).

GTP analogues have a marked effect on Ca^{2+}-channel currents. They clearly slow current activation (Fig. 2*a*; Dolphin & Scott, 1987; Dolphin *et al.*, 1988) and thus appear to inhibit selectively the transient component of the whole-cell current. This is shown by photorelease from an inactive caged precursor of the GTP analogue GTP[S] (guanosine 5′-[γ-thio]triphosphate) inside the cell, which results in gradual inhibition of the Ca^{2+} current, the transient current being completely abolished, whereas the sustained current present at the end of the step is partially reduced (Dolphin *et al.*, 1988) (Fig. 2*a*). However, again this slowed current is not evidence of inhibition of an N current (see below), since it has also been observed in AtT-20 cells which do not possess N channels (Lewis *et al.*, 1986).

Our studies using photorelease of 'caged' GTP analogues show that the Ca^{2+}-channel current which remains available at −30 mV (comprising a largely dihydropyridine-sensitive current) can also be partly inhibited by G-protein activation in DRGs, although the current activation by GTP[S] remains slowed (Fig. 2*b*; Dolphin & Scott, 1989 and A. C. Dolphin & R. H. Scott, unpublished work). Thus, in our hands, GTP[S] and the $GABA_B$ agonist baclofen inhibit both the high-threshold transient current, in agree-

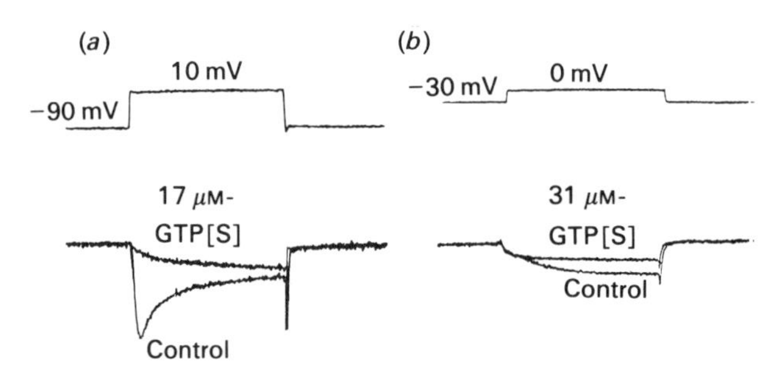

Fig. 2. *Effect of internal GTP[S] on* Ca^{2+}*-channel currents*

(*a*) The larger trace shows the maximim I_{Ba} recorded from V_H − 80 mV in the presence of 100 μM-caged GTP[S]. It was stable for 5 min; subsequently, three flashes were given to release GTP[S] (17 μM), and the smaller current was recorded after 5 min. It was slowly activating and sustained. (*b*) The larger trace shows the maximum I_{Ba} recorded in another cell from V_H − 30 mV. After six flashes to release 31 μM-GTP[S], the smaller trace was recorded after 5 min.

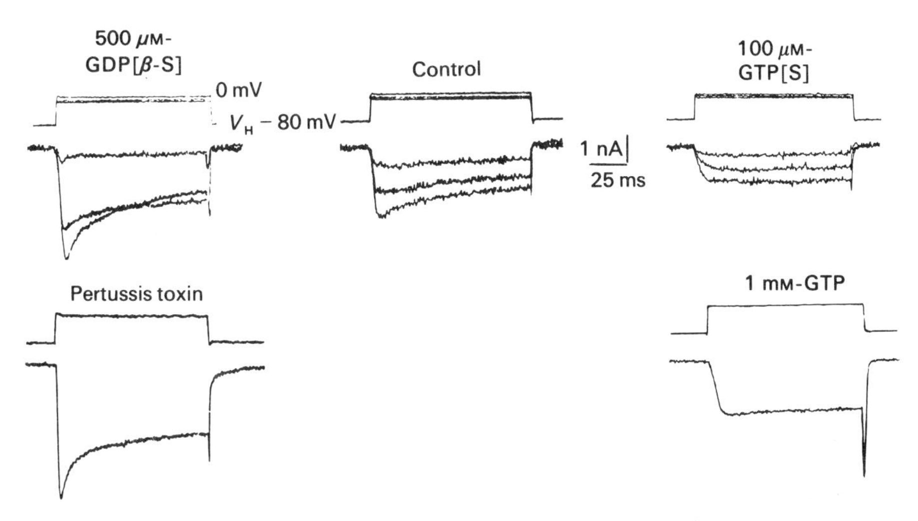

Fig. 3. *Effect of guanine nucleotide analogues and pertussis toxin on Ca^{2+}-channel currents*

A family of control Ca^{2+}-channel currents is shown in the centre panel. For comparison, in the left-hand panel two treatments (internal GDP[β-S] or pretreatment with pertussis toxin), which reduce G-protein activation by endogenous GTP, increase the transient component of I_{Ba}. In the right-hand panel GTP[S] and GTP itself reduced the transient component of I_{Ba}.

ment with Plummer *et al.* (1989), and the high-threshold sustained current, in agreement with Holz *et al.* (1986). The extent of Ca^{2+} current inhibition by neurotransmitters in different cell types must also depend on the receptor–G-protein–channel stoichiometry and to what extent these elements are able to move within the plane of the membrane.

The finding that exogenously added GTP analogues (GTP[S], guanosine 5′-[βγ-imido]triphosphate) inhibit Ca^{2+}-channel currents in many neuronal cells (DRGs, sympathetic neurons, cortical neurons and cerebellar granule cells) led to our suggestion that intracellular levels of endogenous GTP itself might exert a tonic modulatory influence on Ca^{2+} currents (Dolphin & Scott, 1987). In the presence of internal GDP[β-S], which competes with endogenous GTP for binding to the G-protein, the transient component of Ca^{2+} currents is enhanced in amplitude (Fig. 3, left-hand panel) (Dolphin & Scott, 1987). This is also the case in pertussis toxin-treated cells, in which activation of G_i/G_o is prevented by its ADP-ribosylation (Fig. 3, left-hand panel). In addition, high concentrations of GTP (1 mM) have a similar, although not such a marked, effect as GTP[S] on the Ca^{2+}-channel current (Fig. 3, right-hand panel) (Dolphin & Scott, 1989). These findings suggest that there is a tonic partial inhibition of Ca^{2+}-channel currents in DRGs, by a proportion of activated G-proteins.

Voltage Dependence of Ca^{2+}-Channel Current Inhibition

The slowed activation of high-threshold Ca^{2+} currents by GTP analogues, and to a lesser extent by agonists such as (−)-baclofen and GABA, has been suggested to be due to the voltage dependence of the blockade (Grassi & Lux,

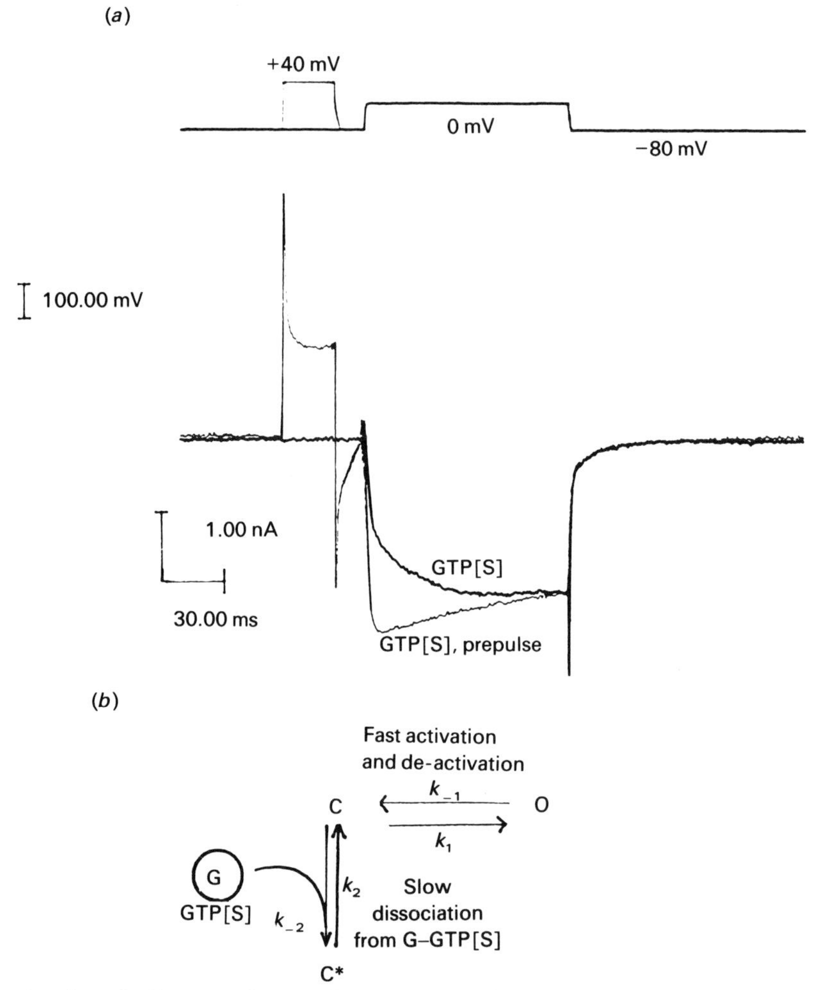

Fig. 4. *Effect of a 20 ms prepulse to +40 mV to increase the rate of activation of I_{Ba} in the presence of GTP[S]*

(*a*) In the presence of 200 μM-internal GTP[S], a test pulse of 100 ms to 0 mV from a holding potential (V_H) of −80 mV activates a slowly activating I_{Ba} (labelled GTP[S]). A 20 ms prepulse to +40 mV, 10 ms before the test pulse, results in activation of a more rapidly activating I_{Ba} (labelled GTP[S], prepulse). (*b*) A model for the interaction between activated G-protein and the closed state of the Ca^{2+}-channel C to produce a modified closed state C*. The rate of conversion of C* to C (k_2) is increased by depolarization as is the rate of conversion of C to O (k_1).

1989; Scott & Dolphin, 1990) (Fig. 4). At hyperpolarized holding potentials there is an interaction (presumably between activated G-protein and a component of the Ca^{2+} channel in the closed state 'C') to form a modified closed state C*, with a rate constant k_{-2} (Fig. 4*b*). C* may be unavailable to open directly. This complex can dissociate by a process with a slow rate constant k_2. Upon dissociation of the activated G-protein from the Ca^{2+} channel, the latter returns to the free closed state C, which is then available to form the open state O with a rapid rate constant k_1 which is dependent on voltage.

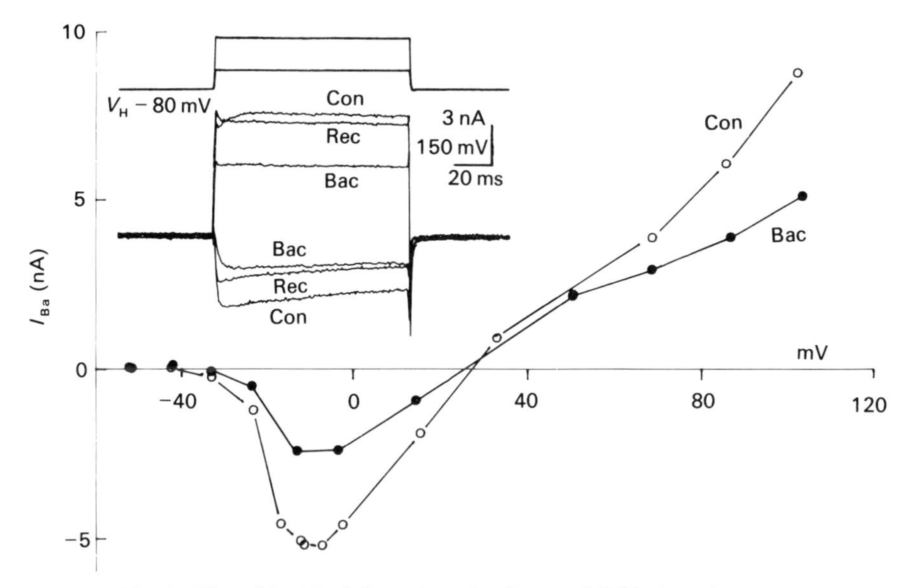

Fig. 5. *Effect of (−)-baclofen on inward and outward Ca^{2+}-channel currents*

Currents were recorded from V_H − 80 mV in a control cell (○) and during application of 100 μM-baclofen (●). The maximum inward current was recorded at 0 mV and the outward current at +100 mV.

Larger depolarizations thus increase the rate of C* → C → O, and, therefore, a large brief depolarizing prepulse applied before the test pulse to 0 mV to activate the maximum inward Ca^{2+} channel current results in the observation of a marked increase in the rate of current activation during the test pulse (Fig. 4*a*). This is because the prepulse has shifted the equilibrium towards O and thus decreased the proportion of C associated with activated G-protein, but, as the interval between the two pulses is brief, re-association to form C* has not occurred. The rate of formation of C* from C (k_{-2}) can be determined by varying the interpulse interval, to be about 0.03 ms^{-1} (Grassi & Lux, 1989). In this analysis it is not essential to assume that the interaction of C with activated G-protein is itself voltage dependent, but only that these channels are not available to open directly, and that O does not associate with activated G-protein.

Although a depolarizing prepulse restores the rate of activation of the Ca^{2+}-channel current to that in the absence of activated G-protein, there is also a non-voltage-dependent reduction in current amplitude (Scott & Dolphin, 1990) which may represent a change in channel phosphorylation state. It is also evident from Fig. 4, that the slowly activating current activated by a test pulse in the presence of GTP[S] has an initial rapidly activating component, representing in this case about 50% of the final current, which is then followed by the slowly activated component. The residual rapidly activating component presumably represents channels which are not associated with activated G-protein (Marchetti & Robello, 1989) and in some cells only a slowly activating component is observed in the presence of GTP[S]. These authors and ourselves (Dolphin & Scott, 1989) found that the time constant of activation of the slowly activating component was not dependent on holding potential, again indicating that the interaction of C with activated G-protein is not voltage dependent.

It has recently been observed by Bean (1989) that noradrenaline did not inhibit frog DRG Ca^{2+}-channel currents activated by depolarizations which were sufficiently large that the net current flowing was outward, owing to Cs^{+} passing out of the cell through Ca^{2+} channels. In contrast, we have observed that outward Ca^{2+}-channel current activated by step depolarizations to about +100 mV are smaller in amplitude in the presence of internal GTP[S] (Dolphin & Scott, 1990). They are also inhibited by (−)-baclofen (Fig. 5). However, whereas the inward current activation is very markedly slowed by agonist and by direct G-protein activation with GTP[S], this effect is much reduced for the outward current, possibly because the rate of formation of O from C is increased at larger step depolarizations and thus the equilibrium between C* αGTP[S] $\rightleftharpoons$ C + αGTP[S] is shifted towards dissociation more rapidly. Again, these results indicate an additional non-voltage-dependent component to neurotransmitter- and G-protein-mediated inhibition of Ca^{2+} currents.

G-Protein Species Involved in Inhibition of Ca^{2+}-Channel Currents

The G-protein involved in inhibition of Ca^{2+}-channel current is clearly pertussis toxin sensitive (Holz *et al.*, 1986; Dolphin & Scott, 1987). Identification of the G-protein involved in coupling receptors to Ca^{2+}-current inhibition has been aided by experiments in which G-proteins are included in the patch pipette when recording Ca^{2+} currents in pertussis toxin-treated cells. It has been found that exogenously added G-proteins can restore the ability of neurotransmitters to inhibit Ca^{2+} currents. These experiments have generally suggested that G_o or its α-subunit is more effective than G_i in restoring coupling (Hescheler *et al.*, 1987; Ewald *et al.*, 1988). In several systems, the use of anti-G-protein antibodies which inhibit function have confirmed that a G_o-protein mediates inhibition of neuronal Ca^{2+} currents by neurotransmitters (Ewald *et al.*, 1988; Harris-Warwick *et al.*, 1988; Brown *et al.*, 1989).

Second Messenger Involvement in the Inhibition of Ca^{2+} Currents

There is no unequivocal evidence that the inhibition of Ca^{2+} channels by neurotransmitter occurs by direct interaction with activated G-protein subunits. However, most evidence suggests that neither activation nor inhibition of adenylate cyclase is essential for the observation of a response (McFazdean & Docherty, 1989; Dolphin *et al.*, 1989*a*). Nevertheless, the activity of L channels (in neurons as in heart) is increased by cyclic AMP-dependent phosphorylation (Gray & Johnson, 1987; Lipscombe *et al.*, 1988; Dolphin *et al.*, 1989*a*). We have recently observed that forskolin is particularly able to increase Ca^{2+} currents by a process involving cyclic AMP-dependent phosphorylation, in the presence of internal GTP[S] (Dolphin, 1990*b*).

The evidence concerning the obligatory mediation of protein kinase C (PKC) in the inhibition of Ca^{2+} currents by neurotransmitters remains equivocal. Rane & Dunlap (1986) initially demonstrated an inhibition of the sustained Ca^{2+} current in chick DRGs by phorbol esters and the synthetic diacylglycerol (DAG) oleoylacetylglycerol (OAG). However, we observed no inhibitory effect of dioctanoyl glycerol on Ca^{2+} channel currents carried by Ba^{2+} (I_{Ba}). Several studies have failed to show either an effect on Ca^{2+} currents of PKC activators or the ability of inhibitors of PKC to prevent the response of Ca^{2+} currents to neurotransmitters or GTP analogues (Wanke *et al.*, 1987; Brezina *et al.*, 1987; Dolphin *et al.*, 1989*a*; McFadzean & Docherty, 1989). The PKC inhibitors polymixin B and H7 both produced some inhibition of Ca^{2+}-channel currents, but did not prevent the effects of guanine nucleotide analogues (Dolphin *et al.*, 1989*a*). A recent study by Hockberger *et al.* (1989) has shown that although OAG inhibits DRG Ca^{2+} currents when applied externally, it had no effect when included intracellularly. This indicates that a direct membrane effect also may contribute to the response to PKC activators.

The neurotransmitters and neuromodulators that inhibit Ca^{2+} currents are not normally considered to be in the class of 'Ca^{2+} mobilizing' agonists that activate phospholipase C. However, we have found (−)-baclofen (100 μM) to increase the production of total inositol phosphates in DRG neurons by 22% in 30 s (Dolphin *et al.*, 1989*a*). For comparison, the enhancement by bradykinin (1 μM) was about 70%. However, we observed bradykinin to have little effect on I_{Ba} or I_{Ca} in most cells. In some cells (7/16) it produced a marked increase of I_{Ba} by $72 \pm 16\%$ (Dolphin *et al.*, 1989*a*; S. M. McGuirk & A. C. Dolphin, unpublished work) (e.g. Fig. 5*b*). In contrast, as previously shown, baclofen produces a substantial inhibition. This suggests that production of neither inositol trisphosphate nor DAG is required for inhibition of Ca^{2+}-channel currents by baclofen.

Inhibition of Low-Threshold Ca^{2+}-Channels

There are contradictory findings on the responsiveness of low-voltage-activated T currents to inhibitory modulation by neurotransmitters or G-protein activation. Gross & MacDonald (1987) showed no effect of dynorphin acting at K receptors, on low-threshold currents in DRGs, although it produced a marked inhibition of the transient high-threshold current. In contrast, Marchetti *et al.* (1986) have shown that, whereas dopamine slows the rate of activation and reduces the amplitude of high-threshold currents in DRG neurons, it inhibits low-threshold currents with no change in their kinetics. Low-threshold single-channel activity recorded from inside-out patches was also inhibited by dopamine. Similarly, we have shown 100 μM-baclofen (Dolphin *et al.*, 1990) and an adenosine agonist, 2-chloroadenosine (Scott & Dolphin, 1987*a*), to inhibit low-threshold Ca^{2+} currents in rat DRGs.

Further evidence in these cells for the association of T channels with a GTP-binding protein comes from our studies showing that photorelease of 10–20 μM-GTP[S] from its inactive caged precursor inhibits T currents in a pertussis toxin-sensitive manner (Dolphin *et al.*, 1989*b*). The time course of inhibition of T currents is fairly slow following liberation of 10 μM-GTP[S],

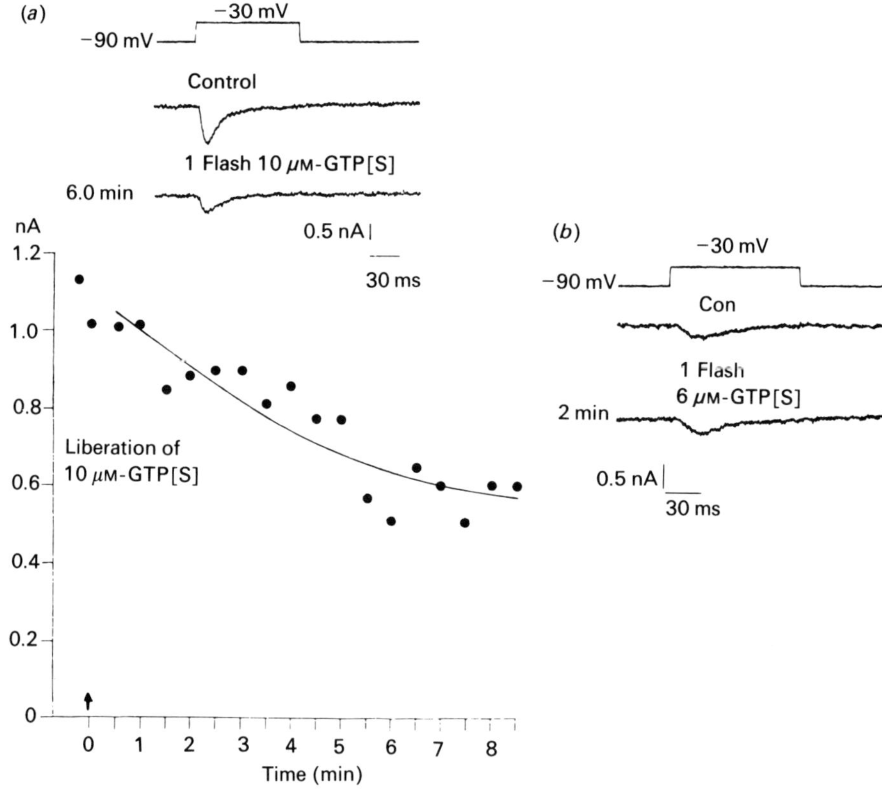

Fig. 6. *Effect of GTP[S] on T currents*

(*a*) 100 μM-caged GTP[S] was included in the patch-pipette solution and allowed to equilibrate with the cell for 10 min before giving a single flash which photolysed the S-caged GTP[S] derivative with 10% efficiency, yielding about 10 μM-free GTP[S] inside the cell. T currents were activated every 30 s at a clamp potential of −30 mV from a holding potential of −90 mV. The T current was stable before GTP[S] liberation at the time indicated by the arrow. The peak amplitude of the T current is plotted against time after photorelease of GTP[S]. The traces show a control T current, and the current recorded 6 min after liberation of 10 μM-GTP[S]. (*b*) Liberation of a lower concentration (6 μM) of GTP[S] from the O-caged precursor with a 6% efficiency produced an increase in T-current amplitude.

and the extent of inhibition is less than 50%. More extensive inhibition of 87 ± 9% was produced by liberation of 17 μM-GTP[S] ($n = 4$).

A recent finding is that photorelease of a lower concentration of GTP[S] (6 μM) consistently increased T currents by 54 ± 20% ($n = 6$) (Fig. 6*b*) (Dolphin *et al.*, 1989*b*). The increase is sensitive neither to pertussis toxin nor to cholera toxin, although both the enhancement by a low concentration of GTP[S] and the inhibition by higher concentrations are prevented by 1 mM-GDP[β-S] (Dolphin *et al.*, 1990). Baclofen also has dual effects on the low-threshold Ca^{2+}-channel current, a low concentration (2 μM) increasing the T current by 25 ± 4% ($n = 6$), whereas 100 μM produced 31 ± 7% inhibition ($n = 5$) (Dolphin *et al.*, 1990). The physiological relevance of neurotransmitter and G-protein-mediated up- and down-modulation of neuronal T currents remains

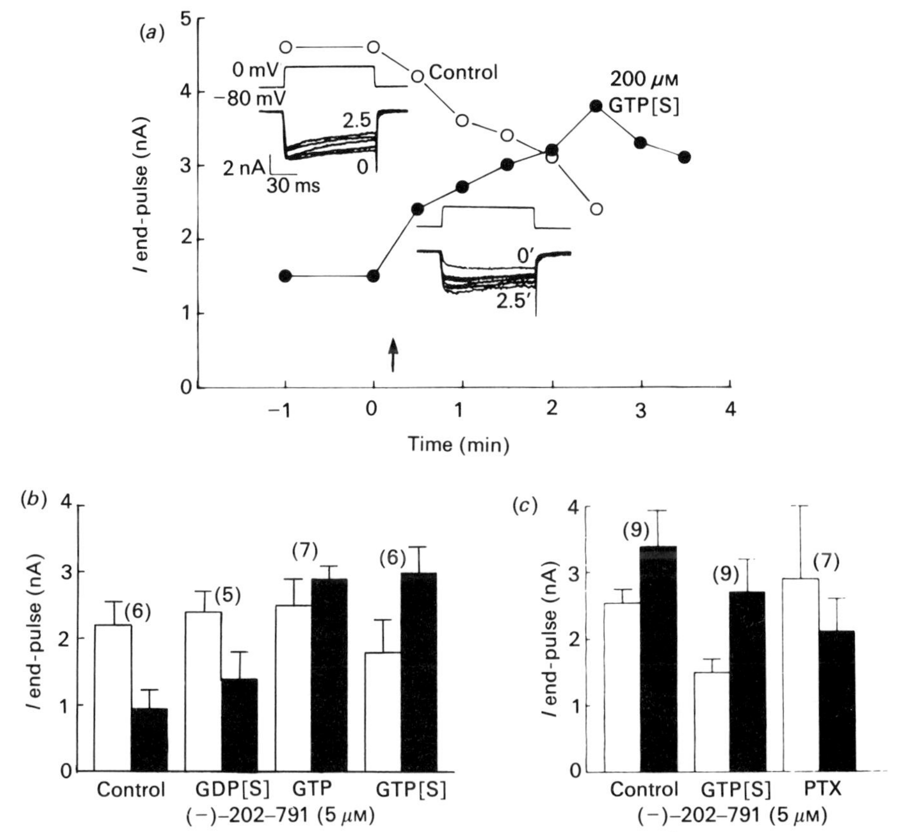

Fig. 7. *Effect of alterations in G-protein activation on the responses of Ca^{2+}-channel currents to the (+) and (−) isomers of (−)-202–791*

(*a*) In a control cell (○) 0.5 μM-(−)-202–791 inhibits I_{Ba}, whereas in a cell containing 200 μM-GTP[S] (●), the same compound shows agonist properties. The application of (−)-202-791 was from a blunt pressure pipette placed near to the cell. Its application was started at the time indicated by the arrow and continued for the rest of the experiment. (*b*) The effect of 5 μM-(−)-202–791 is shown on the maximum I_{Ba} from V_H − 80 mV in control cells, in the presence of 1 mM-internal GDP[β-S], 1 mM-GTP or 200 μM-GTP[S]. The open bars represent the mean I_{Ba} (±S.E.M.) measured at the end of the 100 ms step command. The filled bars represent the maximum effect observed of (−)-202–791, under the different conditions. (*c*) The effect of 5 μM-(+)-202–791 is shown (as in *b*) in control cells, in the presence of GTP[S] and in pertussis toxin (PTX)-treated cells (500 ng/ml, for 2–3 h at 37°C).

to be determined, but these currents are clearly important in regulating patterns of neuronal firing (Llinas, 1988).

Interaction Between Ca^{2+}-Channel Ligand-Binding Sites and G-Protein Activation

Recent work has suggested an interaction between G-protein activation and the effect of Ca^{2+}-channel ligands on L-type Ca^{2+} channels in cultured rat DRGs and sympathetic neurons (Scott & Dolphin, 1987*b*, 1988; Dolphin & Scott, 1989). Ca^{2+}-channel antagonists, including nifedipine (5 μM), (−)-202–791 (0.5–5 μM), diltiazem (30 μM) and D_{600} (10 μM) were observed to

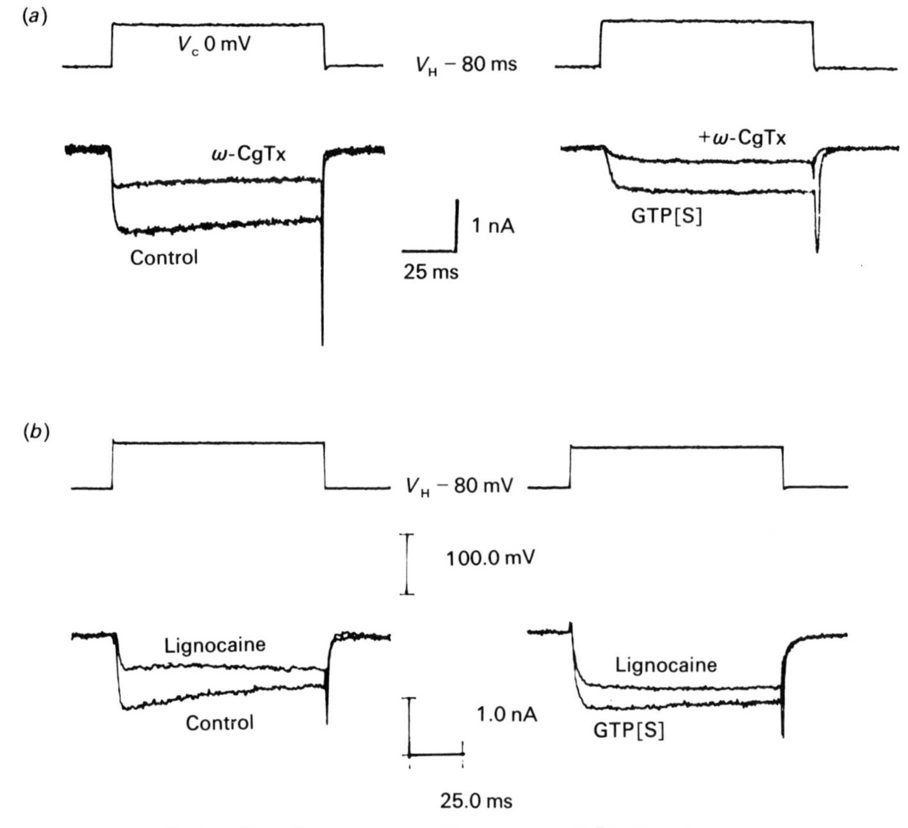

Fig. 8. *Effect of ω-conotoxin and lignocaine on Ca^{2+}-channel currents*

(*a*) ω-Conotoxin (ω-CgTx; 1 μM) inhibited I_{Ba} in control cells (left) and in the presence of internal GTP[S] (right). An initial rapid partial inhibition [70 ± 6% ($n = 11$) in control cells and 58 ± 12% ($n = 5$) in GTP[S]-containing cells] was followed by complete inhibition of I_{Ba} in some cells where application was continued for 5–10 min. (*b*) Lignocaine (100 μM) inhibited I_{Ba} both in control cells (left) and in the presence of internal GTP[S] (right).

show only agonist properties in the presence of internal guanine nucleotide analogues. For example, whereas in control cells 5 μM-nifedipine produced a transient enhancement followed by a persistent inhibition of I_{Ba}, in the presence of internal GTP[S], only enhancement was observed (Scott & Dolphin, 1987*b*) and this was prolonged for the duration of the experiment. This result is not due to an effect on the steady-state inactivation of I_{Ba}, since nifedipine, whether acting as an agonist or an antagonist shifted the steady-state inactivation curve to more hyperpolarized potentials (Dolphin & Scott, 1989). An example of the effect of 0.5 μM-(−)-202–791 on control and GTP[S]-containing neurons is shown in Fig. 7(*a*).

Further studies showed that agonist responses to both 'antagonist' (Scott & Dolphin, 1987*b*) and 'agonist' Ca^{2+}-channel ligands were blocked by pertussis toxin in DRGs (Fig. 7*c*). In addition, the rate of onset of the antagonist response to (−)-202–791 was increased by GDP[β-S] (Dolphin & Scott, 1989). Similarly, internal GTP (1 mM) also induced agonist responses to (−)-202–791, although these were smaller in amplitude than those seen with GTP[S] (Fig. 7*b*). As a control that GTP[S] has not fundamentally affected the pharmacology of the Ca^{2+} channels, the ability of several other Ca^{2+}-channel blockers to inhibit control Ca^{2+}-channel currents and those in the presence of GTP[S] was compared. ω-Conotoxin (1 μM; Fig. 8*a*), ω-agatoxin 1A (10 nM; Adams *et al.*, 1989), Cd^{2+} (100 μM) and lignocaine (100 μM; Fig. 8*b*) inhibited Ca^{2+}-channel currents similarly under the two conditions.

These findings indicate that in the intact neuronal cells studied, a G-protein which can be activated by GTP or its non-hydrolysable analogues is an essential requirement for Ca^{2+}-channel ligands to act as agonists. Purified dihydropyridine receptors have Ca^{2+}-channel activity when inserted in lipid bilayers, and show an agonist response to Bay K8644 (Hymel *et al.*, 1988), but it has also been observed that additional Ca^{2+}-channel agonist responses may result from Ca^{2+}-channel phosphorylation or inhibition of dephosphorylation (Armstrong & Eckert, 1987; Lory *et al.*, 1990). We have recently found that the agonist response to (−)-202–791 resembles the effect of cyclic AMP-dependent phosphorylation by forskolin in GTP[S]-containing cells (Dolphin, 1990*b*). Thus, in the presence of GTP[S], (−)-202–791 binds to the Ca^{2+}-channel and allows the phosphorylation state of the channel to be increased, thus increasing the macroscopic current. It remains unclear how interaction of the channel with activated pertussis toxin-sensitive G-proteins modifies either Ca^{2+}-channel ligand binding (Horne *et al.*, 1988) or cyclic AMP-dependent phosphorylation.

Most binding studies have shown equivocal effects of guanine nucleotide analogues on Ca^{2+}-channel ligand binding; however, a recent report has found that GTP[S] enhanced the binding of Bay K8644 to rat brain membranes in a pertussis toxin-sensitive manner (Bergamaschi *et al.*, 1988). Clearly, the interaction between Ca^{2+} channels and G-proteins is complex, and is likely to be unravelled by reconstitution experiments with pure receptors, channels and G-proteins.

Which Types of Ca^{2+} Channel Mediate Transmitter Release?

The release of neurotransmitters from presynaptic terminals is dependent on an influx of Ca^{2+} through voltage-sensitive channels. The subtypes of Ca^{2+} channel involved in transmitter release remain equivocal. It has been suggested that N channels are the most important (Hirning *et al.*, 1988; Tsien *et al.*, 1988), primarily because transmitter release is not markedly sensitive to dihydropyridine Ca^{2+}-channel antagonists, although it is sensitive to agonists (Rane *et al.*, 1987). However, dihydropyridine antagonists show a very marked voltage sensitivity and are poorly effective at hyperpolarized membrane potentials (Sanguinetti & Kass, 1984). Indeed, we have recently observed that when cultured cerebellar neurons are maintained in a depolarized state in 50 mM-K^+, Ca^{2+}-free medium, and glutamate release is stimulated by Ca^{2+}, this release is almost completely inhibited by dihydropyridine antagonists (Fig. 9*a*), the concentration required to inhibit by 50% (IC_{50}) for (−)-202–791 being about 1 nM (Huston & Dolphin, 1990). Thus these results indicate that

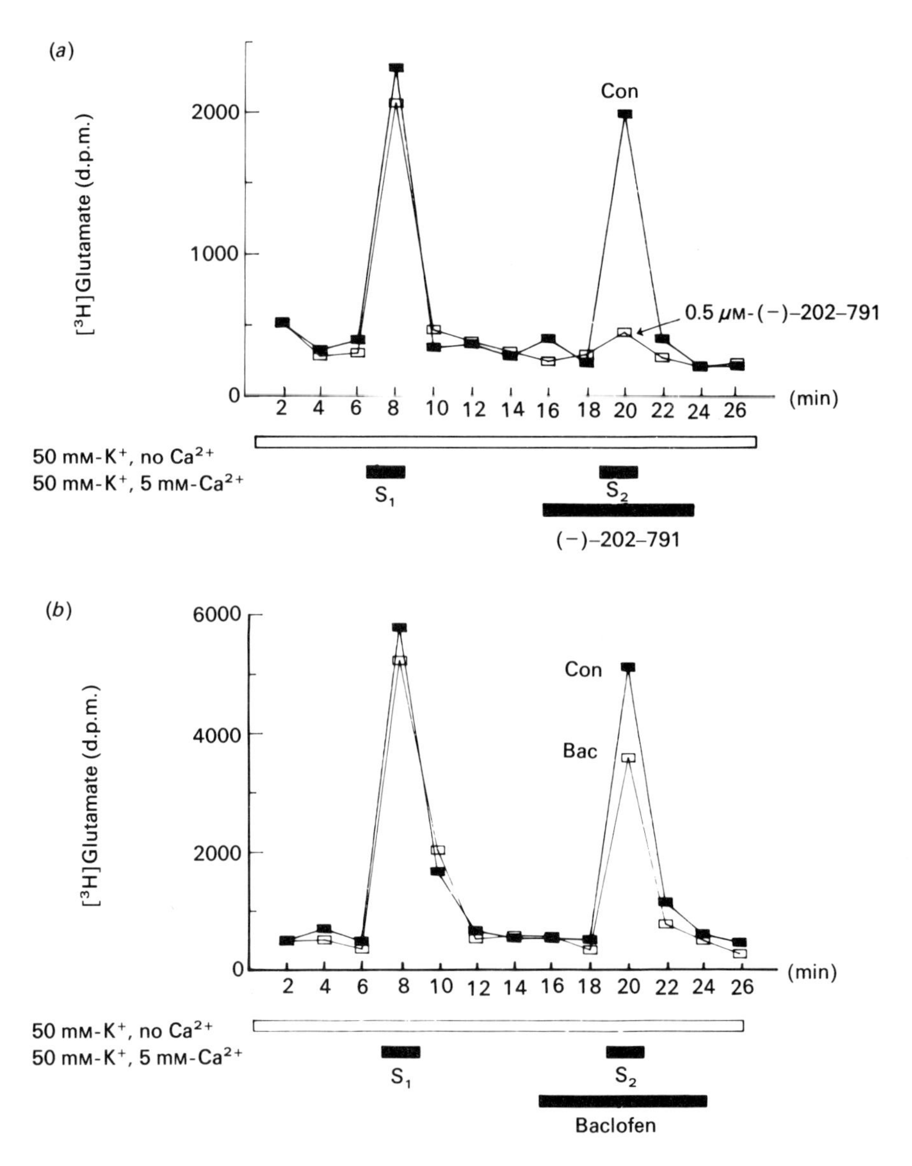

Fig. 9. *Effect of (−)-202–791 and baclofen on glutamate release from depolarized cerebellar granule neurons*

Granule neurons were grown on coverslips for 7–15 days and glutamate release was measured by preincubation of the neurons with [^{3}H]glutamate as previously described (Dolphin & Prestwich, 1985). Neurons were then maintained in a depolarizing medium containing 50 mM-K^+ in the absence of Ca^{2+}. Glutamate release was then stimulated by 2 min incubation in a medium containing 50 mM-K^+ together with 5 mM-Ca^{2+}. Drugs were included before and during S_2. (*a*) (−)-202–791 (0.5 μM) almost completely prevented glutamate release in S_2. (*b*) Baclofen (100 μM) produced a 30% reduction of release in S_2.

dihydropyridine-sensitive Ca^{2+} channels are able to subserve transmitter release. It remains unclear whether release is normally poorly sensitive to dihydropyridine antagonists because it is subserved by N channels, which are unaffected by dihydropyridines, or whether L channels are involved, but show little response to dihydropyridines during brief depolarizations.

The other agent that has been used to dissect out the channels involved in transmitter release is ω-conotoxin. This inhibits transmitter release in many, but by no means all, systems, but there is disagreement as to whether it selectively inhibits N channels (Plummer *et al.*, 1989), or whether L channels are also affected (McCleskey *et al.*, 1987). Thus its diagnostic use remains limited.

A further argument for the prime importance of N channels in the mediation of transmitter release is that 'N current' can be inhibited by various neurotransmitters that also subserve presynaptic inhibition. However, as discussed above, there is also evidence that inhibition of the transient component of whole-cell Ca^{2+} current is due to a slow recovery from voltage-dependent block of the current, and that L current can be inhibited by neurotransmitters. In addition, in recent experiments we have shown that (−)-baclofen is capable of inhibiting neurotransmitter release evoked by Ca^{2+} from cerebellar neurons maintained in a depolarized state, where the Ca^{2+} channels involved are entirely dihydropyridine sensitive (Fig. 9*b*) (Huston & Dolphin, 1990). Thus the involvement of different Ca^{2+}-channel types in physiological synaptic transmission and its inhibition by neurotransmitters remains an open question.

Conclusion

We have outlined the evidence for direct and indirect modulation of Ca^{2+}-channel activity by G-proteins. Ca^{2+} currents may be inhibited by activation of a family of neurotransmitter receptors ($GABA_B$, adenosine A_1, noradrenaline α_2, etc.) by activation of pertussis toxin-sensitive G-proteins. Presynaptic inhibition by the same family of neurotransmitters and modulators may result from both activation of K^+ currents and inhibition of Ca^{2+} currents. However, many questions remain to be answered concerning the mechanisms responsible for this form of inhibition, and the subtypes of Ca^{2+}-channel involved.

References

Adams, M. E., Bindokas, V. P., Dolphin, A. C. & Scott, R. H. (1989) *J. Physiol. (London)* **418**, 34*P*
Armstrong, D. & Eckert, R. (1987) *Proc. Natl. Acad. Sci. U.S.A.* **84**, 2518–2522
Augustine, G. J., Charlton, M. P. & Smith, S. J. (1987) *Annu. Rev. Neurosci.* **10**, 633–693
Bean, B. P. (1989) *Nature (London)* **340**, 153–156
Bergamaschi, S., Govoni, S., Cominetti, P., Parenti, M. & Trabucchi, M. (1988) *Biochem. Biophys. Res. Commun.* **56**, 1279–1286
Brezina, V., Eckert, R. & Erxleben, C. (1987) *J. Physiol. (London)* **388**, 565–595
Brown, D. A., McFadzean, I. & Milligan, G. (1989) *J. Physiol. (London)* **415**, 20*P*
Burlhis, T. M. & Aghajanian, G. K. (1987) *Synapse* **1**, 582–588
Carbone, E. & Lux, H. D. (1984) *Nature (London)* **310**, 501–502
Choi, D. W. (1988) *Trends Neurosci.* **11**, 465–469
Coulter, D. A., Huguenard, J. R. & Prince, D. A. (1989) *Neurosci. Lett.* **98**, 74–78

Docherty, R. J. & McFadzean, I. (1989) *Eur. J. Neurosci.* **1**, 135–140
Dolphin, A. C. (1987) *Trends Neurosci.* **10**, 53–57
Dolphin, A. C. (1990*a*) *Annu. Rev. Physiol.* **52**, 243–255
Dolphin, A. C. (1990*b*) *J. Physiol.* in the press
Dolphin, A. C. & Prestwich, S. A. (1985) *Nature* (*London*) **316**, 148–150
Dolphin, A. C. & Scott, R. H. (1987) *J. Physiol.* (*London*) **386**, 1–17
Dolphin, A. C. & Scott, R. H. (1989) *J. Physiol.* (*London*) **413**, 271–288
Dolphin, A. C. & Scott, R. H. (1990) *Eur. J. Neuroscience* **2**, 104–108
Dolphin, A. C., Wootton, J. F., Scott, R. H. & Trentham, D. R. (1988) *Pflügers Arch.* **411**, 628–636
Dolphin, A. C., McGuirk, S. M. & Scott, R. H. (1989*a*) *Br. J. Pharmacol.* **97**, 263–273
Dolphin, A. C., Scott, R. H. & Wootton, J. F. (1989*b*) *J. Physiol.* (*London*) **410**, 16*P*
Dolphin, A. C., Scott, R. H. & Wootton, J. F. (1990) *J. Physiol.* (*London*) **423**, 89P
Dunlap, K. & Fischbach, G. (1981) *J. Physiol.* (*London*) **317**, 519–535
Eckert, R. & Tillotson, D. L. (1981) *J. Physiol.* (*London*) **314**, 265–280
Ewald, D. A., Sternweis, P. C. & Miller, R. J. (1988) *Proc. Natl. Acad. Sci. U.S.A.* **85**, 3633–3637
Gilman, A. G. (1987) *Annu. Rev. Biochem.* **56**, 615–649
Gray, R. & Johnson, D. (1987) *Nature* (*London*) **327**, 620–622
Grassi, F. & Lux, H. D. (1989) *Neurosci. Lett.* **105**, 113–117
Gross, R. A. & MacDonald, R. L. (1987) *Proc. Natl. Acad. Sci. U.S.A.* **84**, 5469–5473
Harris-Warwick, R. M., Hammond, C., Paupardin-Tritsch, D., Homburger, V., Rouot, B., Bockaert, J. & Gerschenfeld, H. M. (1988) *Neuron* **1**, 27–32
Hescheler, J., Rosenthal, W., Trautwein, W. & Schultz, G. (1987) *Nature* (*London*) **325**, 445–447
Hirning, D., Fox, A. P., McCleskey, E. W., Olivera, B. M., Thayer, S. A., Miller, R. J. & Tsien, R. W. (1988) *Science* **239**, 57–61
Hockberger, P., Toselli, M., Swandulla, D. & Lux, H. D. (1989) *Nature* (*London*) **338**, 340–342
Holz, G. G., Rane, S. G. & Dunlap, K. (1986) *Nature* (*London*) **319**, 670–672
Horne, W. A., Abdel-Ghany, M., Racker, E., Wieland, G. A., Oswald, R. E. & Cerione, R. A. (1988) *Proc. Natl. Acad. Sci. U.S.A.* **85**, 3718–3722
Huston, E. & Dolphin, A. C. (1990) *Br. J. Pharmacol.* **99**, 129P
Hymel, L., Streissnig, J., Glossmann, H. & Schindler, H. (1988) *Proc. Natl. Acad. Sci. U.S.A.* **85**, 4290–4294
Lewis, D. L., Weight, F. F. & Luini, A. (1986) *Proc. Natl. Acad. Sci. U.S.A.* **83**, 9035–9039
Lipscombe, D., Bley, K. & Tsien, R. W. (1988) *Soc. Neurosci. Abstr.* **14**, 64.12
Llinas, R. (1988) *Science* **242**, 1654–1664
Llinas, R. & Yarom, Y. (1981) *J. Physiol.* (*London*) **315**, 569–584
Lory, P., Nargeot, J., Richard, S. & Tiaho, F. (1990) *J. Physiol.* **418**, 24*P*
Marchetti, C. & Robello, M. (1989) *Biophys. J.* **56**, 1267–1272
Marchetti, C., Carbone, E. & Lux, H. D. (1986) *Pflügers Arch.* **406**, 104–111
McCleskey, E. W., Fox, A. P., Feldman, D., Cruz, L. Y., Olivera, B. M., Tsien, R. W. & Yoshikami, D. (1987) *Proc. Natl. Acad. Sci. U.S.A.* **84**, 4327–4331
McFadzean, I. & Docherty, R. J. (1989) *Eur. J. Neurosci.* **1**, 141–147
Nowycky, M. C., Fox, A. P. & Tsien, R. W. (1985) *Nature* (*London*) **316**, 440–443
Plummer, M. R., Logothetis, D. E. & Hess, P. (1989) *Neuron* **1**, 1453–1463
Rane, S. G. & Dunlap, K. (1986) *Proc. Natl. Acad. Sci. U.S.A.* **83**, 184–188
Rane, S. G., Holz, G. G. & Dunlap, K. (1987) *Pflügers Archiv.* **409**, 361–366
Sanguinetti, M. C. & Kass, R. S. (1984) *Circ. Res.* **55**, 336–348
Scott, R. H. & Dolphin, A. C. (1986) *Neurosci. Lett.* **56**, 59–64
Scott, R. H. & Dolphin, A. C. (1987*a*) in *Topics and Perspectives in Adenosine Research* (Gerlach, E. & Becker, B. F., eds.), pp. 549–558, Springer-Verlag, Berlin
Scott, R. H. & Dolphin, A. C. (1987*b*) *Nature* (*London*) **330**, 760–762
Scott, R. H. & Dolphin, A. C. (1988) *Neurosci. Lett.* **89**, 170–175
Scott, R. H. & Dolphin, A. C. (1990) *Br. J. Pharmacol.* **99**, 629–630
Swandulla, D. & Armstrong, C. M. (1988) *J. Gen. Physiol.* **92**, 197–218
Tsien, R. W., Lipscombe, D., Madison, D. V., Bley, K. R. & Fox, A. P. (1988) *Trends Neurosci.* **11**, 431–437
Wanke, E., Ferroni, A., Malgaroli, A., Ambrosini, A., Pozzan, T. & Meldolesi, J. (1987) *Proc. Natl. Acad. Sci. U.S.A.* **84**, 4313–4317

Biochem. Soc. Symp. **56**, 61–69
Printed in Great Britain

Antibodies as Probes of G-Protein Receptor–Effector Coupling and of G-Protein Membrane Attachment

ALLEN M. SPIEGEL, WILLIAM F. SIMONDS, TERESA L. Z. JONES,
PAUL K. GOLDSMITH and CECILIA G. UNSON*

*Molecular Pathophysiology Branch, National Institute of Diabetes, Digestive and Kidney Diseases, National Institutes of Health, Bethesda, MD 20892, U.S.A. and *Department of Biochemistry, Rockefeller University, New York, NY 10021, U.S.A.*

Synopsis

The specificity of antisera raised against synthetic decapeptides corresponding to the *C*-terminus of G-protein α-subunits was rigorously defined. Antisera raised against α-subunit *C*-terminal decapeptides proved capable of immunoprecipitating their cognate G-proteins, as well as recognizing these proteins in native cell membranes. Thus the α_s-specific antiserum could block agonist-stimulated adenylyl cyclase activity in native membranes and also immunoprecipitate an activated α_s–adenylyl cyclase complex. The α_i2-, but not α_i3- and α_z-specific antiserum could block agonist-mediated inhibition of adenylyl cyclase in human platelet membranes. These results indicate that the *C*-terminal decapeptide is involved in G-protein receptor, but not effector, coupling. These antisera also proved useful in immunoprecipitation of endogenous and transfected α-subunits in COS cells. Using this approach, we were able to show that both α_s and α_i are membrane associated, but only the latter is myristylated. A mutant α_i1 (second residue Gly changed to Ala) fails to incorporate myristate and is localized in the soluble fraction. Myristylation is thus essential for membrane attachment of α_i, but not α_s.

Introduction

G-proteins involved in signal transduction are members of a guanine-nucleotide-binding protein superfamily that includes cytoskeletal proteins such as tubulin, soluble proteins (initiation and elongation factors involved in protein synthesis), and low molecular mass GTP-binding proteins such as the *ras* p21 proto-oncogenes and *ras*-related proteins [1–3]. Members of the G-protein subset of the GTP-binding protein superfamily share certain general features in common with other members of the GTP-binding protein superfamily: (i) all GTP-binding proteins bind guanine nucleotides with high affinity and specificity, and possess intrinsic GTPase activity that modulates interactions between the GTP-binding protein and other elements, and (ii) GTP-binding proteins serve as substrates for ADP-ribosylation by bacterial toxins; this covalent modification disrupts normal function.

G-proteins share other features that distinguish them from other GTP-binding proteins. These features include their: (i) association with the cytoplasmic surface of the plasma membrane (*ras* p21 and some other low molecular mass GTP-binding proteins are also associated with the cytoplasmic membrane surface); (ii) function as receptor–effector couplers and (iii) heterotrimeric structure. G-proteins contain α-, β-, and γ-subunits, each distinct gene products. The latter two subunits are tightly, but non-covalently, linked in a βγ complex. α-Subunits bind guanine nucleotide, serve as toxin substrates, confer specificity in receptor–effector coupling and directly modulate effector activity. Upon activation by GTP, α-subunits are thought to dissociate from the βγ-complex. The βγ-complex is required for G-protein–receptor interaction, can inhibit G-protein activation by blocking α-subunit dissociation and may, in some cases, directly regulate effector activity.

Molecular cloning provides evidence for a minimum of nine distinct α-subunit genes, including G_s, G_o, G_i1, G_i2, G_i3, G_t1, G_t2 [1–3], $G_{x(z)}$ [4–6] and G_{olf} [7]. There may be additional α-subunit genes [8]. Further diversity is created by alternative splicing leading to the expression of four forms of G_s [9]. At least two distinct genes each exist for both β- and γ-subunits. The expression of certain α-subunits is highly restricted, e.g. G_t1 and G_t2 are found only in photoreceptor rod and cone cells, respectively, and G_{olf} only in olfactory neurons, whereas others such as G_s are expressed ubiquitously.

The specificity of G-protein interactions with receptors and effectors has been defined in very few cases. Studies involving reconstitution of purified receptors and G-proteins in phospholipid vesicles showed that β-adrenergic receptors couple, in decreasing order of efficiency, to $G_s > G_i \gg G_t$, and that for rhodopsin, the selectivity of coupling is $G_t = G_i \gg G_s$ [10]. Similar studies involving G-protein–effector interaction indicated that only G_s can activate adenylyl cyclase (G_s also appears to stimulate another effector, a Ca^{2+} channel), and that G_t uniquely activates retinal cyclic GMP phosphodiesterase (see [11], for example). The endogenous G-proteins coupled to most other receptors and effectors, however, remain to be identified. Many G-protein-coupled receptors (D_2-dopaminergic; for example, see [12]), and a variety of effectors, including adenylyl cyclase (inhibition), certain Ca^{2+} (inhibition) and K^+ (stimulation) channels, and phospholipase C (stimulation) in cell types such as neutrophils (e.g. by fMet-Leu-Phe receptor) are regulated by one or more pertussis toxin-sensitive G-proteins [1–3]. Since, not only both forms of G_t, but also G_i1, G_i2, G_i3 and G_o, are pertussis toxin-sensitive G-proteins, demonstration of an effect of pertussis toxin on receptor or effector regulation does not uniquely identify the relevant endogenous G-protein. In most cells, phospholipase C is regulated by a pertussis toxin-insensitive G-protein. $G_{z(x)}$ may play this role, but definitive evidence is lacking.

Generation and Characterization of G-Protein Peptide Antisera

We have generated numerous antisera against synthetic peptides corresponding to the predicted amino acid sequence of G-protein subunits. Such antisera have proved very useful in identifying the putative proteins encoded by cloned cDNAs [13,14], in discriminating between closely related G-proteins [15] and in quantifying and localizing G-proteins. For example, three antisera, LD, LE, and SQ, were raised against an internal decapeptide (residues 159–168) of α-subunits of G_i1, G_i2, and G_i3, respectively. These proved capable of discriminating between α-subunits of 41 and 40 kDa purified from brain, and of 41 kDa purified from HL-60 cells, and served to identify these as the products of G_i1, G_i2, and G_i3, respectively [13,15].

One approach theoretically capable of identifying endogenous G-proteins coupling to particular receptors and effectors involves the use of specific antibodies that can bind to native G-proteins and affect their function. Antibodies directed against the *C*-terminus of G_α-subunits hold particular promise in this regard. Several lines of evidence point to the importance of the extreme *C*-terminus in G-protein–receptor coupling. These include the locus of the receptor-uncoupling mutation (*unc*) in S49 mouse lymphoma cells (sixth residue from the *C*-terminus) [16] and the ability of pertussis toxin-catalysed ADP-ribosylation of a cysteine residue (fourth from the *C*-terminus) to uncouple G-proteins from receptors [17].

To study G-protein coupling with this approach, we generated antisera capable of recognizing each of the known G_α-subunits (Table 1). We initially assessed the reactivity and specificity of these antibodies against the immunogenic and related peptides by enzyme-linked immunoabsorbent assay (ELISA). The results indicated that antisera RM and QN are highly specific for their cognate peptides, whereas AS, EC and GO show substantial cross-reactivity for each other's immunogenic peptide. This presumably reflects the relatively unique sequences of the RM and QN peptides versus the more homologous sequences of the AS, EC and GO peptides (Table 1). We next tested the reactivity of the antisera with proteins encompassing the immunogenic peptide sequences by performing the immunoblots and immunoprecipitation of bacterial lysates containing unique, defined G_α subunits expressed by recombinant DNA techniques. On immunoblot, RM and QN antisera are absolutely specific for G_s and G_x, respectively. G_s and G_x, moreover, are not reactive with GO, EC or AS antisera. AS antiserum reacts strongly and equivalently with G_i1 and G_i2, shows weak reactivity against G_i3, and no reactivity against G_o. GO and EC antisera react best with G_o and G_i3, respectively, but they display

Table 1. *C-Terminal decapeptides and corresponding antisera*

Peptide	Sequence*	Antiserum†	G-α-subunit‡
RM	RMHLRQYELL	RM	G_s
QN	QNNLKYIGLC	QN	$G_{x(z)}$
GO	ANNLRGCGL*Y*	GO	G_o
EC	KNNLKECGL*Y*	EC	G_i3
KE	KENLKDCGL*F*	AS	G_t1, G_t2
KN	KNNLKDCGL*F*	—	G_i1, G_i2

* Single-letter amino acid code.
† Antiserum raised by immunization with designated peptide.
‡ Designated peptide represents *C*-terminus of corresponding G-α-subunit [5].

very substantial reciprocal cross-reactivity. The somewhat unexpected pattern of reactivity of AS, GO and EC reflects the importance of the *C*-terminal residue, phenylalanine versus tyrosine, in determining antigenicity (Table 1). The identical pattern of reactivity was observed in immunoprecipitation experiments which also indicated that the antisera can recognize native G-proteins.

Peptide Antibodies as Probes of G-Protein Receptor–Effector Coupling

Using purified proteins reconstituted into phospholipid vesicles, we found that affinity-purified AS antibodies specifically block rhodopsin–G_t (transducin) interaction, but do not block interaction between activated G_t and its effector, cyclic GMP phosphodiesterase [18]. This approach was then extended to G-proteins *in situ*, i.e. in native membranes [19]. RM antibodies were shown on immunoblots to recognize the multiple forms of $G_s\alpha$ derived from alternative splicing. Affinity-purified RM antibodies effectively block receptor-mediated adenylyl cyclase stimulation by the β-adrenergic agonist, isoprenaline, in S49 mouse lymphoma cell membranes. Fluoride stimulation (which bypasses the receptor to activate G_s) is only partially blocked by RM antibodies at much higher concentrations than required for inhibition of isoprenaline-stimulated activity (Fig. 1). Antibody inhibition of adenylyl cyclase stimulation could be completely prevented by the cognate peptide. When membranes are preactivated with fluoride or with guanosine 5′-[γ-thio]-triphosphate, RM antibodies specifically immunoprecipitate an active G_s–adenylyl cyclase complex.

In human platelet membranes, we identified G_s, G_i2, G_i3, and $G_{x(z)}$, but neither G_i1 nor G_o, by immunoblot. As for S49 lymphoma cell membranes, RM antibodies, by binding to α_s, could block stimulation of adenylyl cyclase

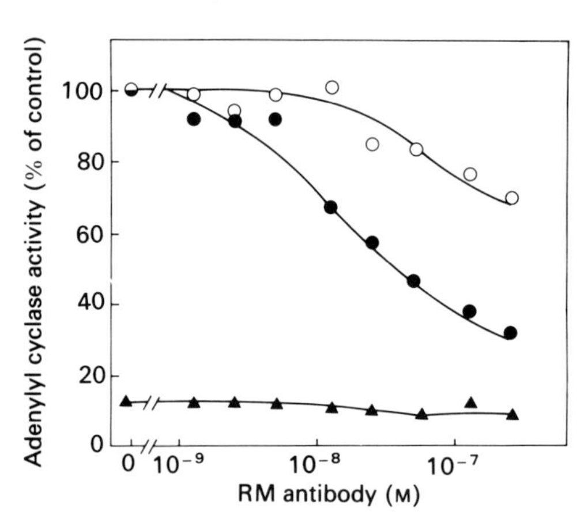

Fig. 1. *RM antibody inhibition of* G_s *activation in S49 mouse lymphoma cell membranes*

Membranes were incubated with increasing concentrations (as shown) of RM affinity-purified antibody for 2 h at 4°C, after which aliquots were assayed for adenylyl cyclase activity with 10 mM-NaF (○), or with 100 μM-GTP with (●) or without (▲) 500 μM-*l*-isoprenaline (as shown). Values (in pmol of cyclic AMP/min per mg) are the means of triplicate determinations and are expressed as a percentage of the control (84 for NaF and 60 for isoprenaline). (From [19].)

by agonist [prostaglandin E_1 (PGE_1) in platelet membranes]. AS, but not EC or QN, antibodies could block inhibition of adenylyl cyclase by the α_2-adrenergic agonist noradrenaline (Fig. 2). Since we could not identify G_i1 in platelet membranes, and antibodies EC and QN, which bind to G_i3, and $G_{x(z)}$, respectively, do not block adenylyl cyclase inhibition, we conclude that G_i2 can couple to the α_2-adrenergic receptor and thereby mediate adenylyl cyclase inhibition [20]. This provides the first evidence in native membranes for interaction between a specific G-protein and receptor–effector.

These results have major implications for our understanding of G-protein structure and function and for the elucidation of the specificity of receptor–effector coupling by G-proteins. The ability of *C*-terminal decapeptide antibodies to block receptor–G-protein interaction emphasizes the importance of this domain of the α-subunit in receptor interaction. The ability of antibodies to interact with the membrane-bound G-protein implies that this region is exposed at the cytoplasmic surface and is consistent with the ability of pertussis toxin to covalently modify a cysteine within the *C*-terminal decapeptide in the membrane-bound G-protein. Antibody binding to the *C*-terminus, however, does not result in global inhibition of G-protein function. Indeed, RM antibody immunoprecipitation of a G-protein–effector complex implies that the *C*-terminal decapeptide of the α-subunit is not critically involved in effector interaction.

The ability of *C*-terminal decapeptide antibodies to block receptor–G-protein interaction in native membranes should be useful in defining the specificity of these interactions. The studies described above indicate that, at a minimum, one should be able to distinguish coupling to G_i1 and/or G_i2 from coupling to either G_i3 or to G_o. Studies are in progress to extend this approach to other tissues and cells with varying proportions of pertussis toxin-sensitive G-proteins. Likewise, immunoprecipitation of activated G-protein–effector complexes by specific *C*-terminal peptide antibodies should

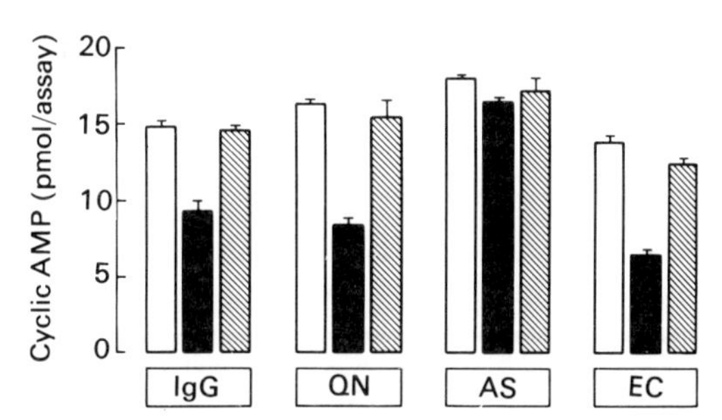

Fig. 2. *Effect of C-terminal peptide antisera on α_2-adrenergic inhibition of adenylyl cyclase in human platelet membranes*

Aliquots of membranes were incubated with control rabbit immunoglobulin (IgG), or affinity-purified antibodies QN, AS or EC at a final antibody concentration of 50 μg/ml. After 1 h at 4°C, PGE_1-stimulated adenylyl cyclase activity was determined with or without noradrenaline and yohimbine: □, 10 μM-PGE_1; ■, 50 μM-noradrenaline; ▧, 50 μM-noradrenaline + 50 μM-yohimbine. Values are the means ± S.E.M of triplicate determinations. (From [20].)

prove helpful in identifying effectors linked to specific G-proteins. Recombinant DNA techniques will clearly provide important information on receptor–effector coupling by G-proteins, e.g. α-subunits of G_i1, G_i2, and G_i3 expressed in *Escherichia coli* all proved capable of stimulating a K^+ channel [21], but such approaches can only indicate which G-proteins are potentially capable of coupling to individual receptors and effectors. The use of specific antibodies offers the possibility of identifying the endogenous G-protein(s) that actually couple to individual receptors and effectors.

Peptide Antibodies as Probes of G-Protein Membrane Attachment

Both G-protein α- and βγ-subunits are tightly attached to the cytoplasmic surface of the plasma membrane [1,2]. Even after activation (and presumed subunit dissociation), neither α- nor βγ-subunits are released from the membrane with aqueous buffers lacking detergents [22]. The molecular basis for the tight association of G-protein subunits with the plasma membrane is not apparent from inspection of the primary sequences predicted by cloned cDNAs [1,2,5]. None of the subunits shows stretches of hydrophobic residues predicted to span the plasma membrane. Hydrodynamic studies of G-protein subunits in buffers with and without detergent have led to the suggestion that α-subunits are intrinsically hydrophilic, whereas βγ-subunits are hydrophobic and require detergent to prevent aggregation. The failure of α-subunits, moreover, to associate with phospholipid vesicles unless these contained βγ-subunits led to the suggestion that the latter serve to anchor α-subunits in the plasma membrane [23].

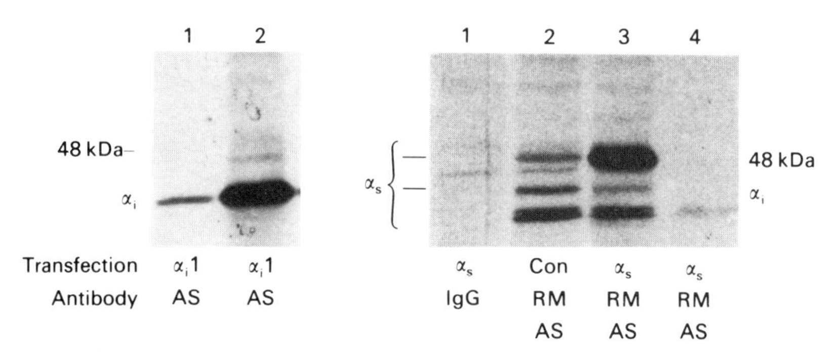

Fig. 3. *Immunoprecipitation of G-protein α-subunits in transfected COS cells*

COS cells were transfected using the diethylaminoethyldextran method with the cDNA for α_s or α_i1 inserted into the pCD-PS expression vector or sham-transfected (Con) and radiolabelled after 48 h with [^{35}S]methionine or [^{3}H]myristic acid. A particulate fraction was prepared after cell lysis, homogenization, low-speed centrifugation (1000 *g* for 3 min) and ultra-centrifugation of the supernatant (400 000 *g* for 30 min). Equivalent amounts of protein were immunoprecipitated with rabbit IgG, affinity-purified RM antibodies specific for the α_s-subunit or affinity-purified AS antibodies specific for α_i1 and α_i2. Samples were analysed by SDS/PAGE, fixed, treated with En^{3}Hance (Dupont) and exposed to XAR-2 film at −70°C for 7–21 days. An endogenous [^{35}S]methionine-radiolabelled protein of molecular mass 48 kDa, determined by molecular mass markers, is labelled for reference in the Figure. Left panel: [^{3}H]myristic acid (lane 1) and [^{35}S]methionine (lane 2)-labelled proteins (13 μg of protein). Right panel: [^{35}S]methionine (lanes 1–3) and [^{3}H]myristic acid (lane 4)-labelled proteins (13 μg of protein). (From [26].)

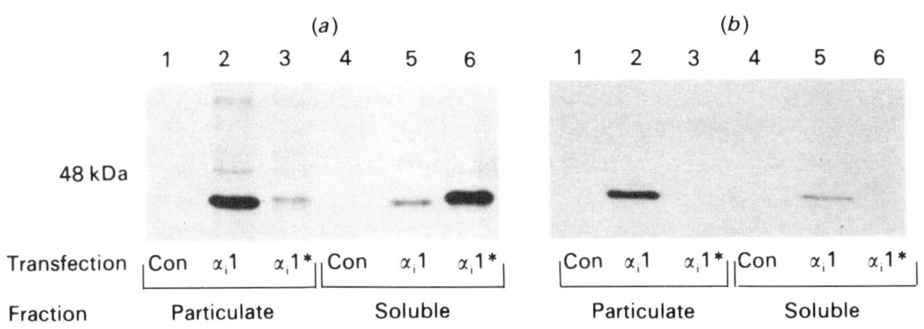

Fig. 4. *Immunoprecipitation of cellular fractions after transfection with the mutant* $\alpha_i 1$

COS cells were transfected with the pCD-PS vector without inserts (Con) or with cDNA for the wild-type $\alpha_i 1$ or the mutant $\alpha_i 1(\alpha_i 1^*)$ and radiolabelled after 48 h with [^{35}S]methionine or [^{3}H]myristic acid. The cells were lysed, homogenized and centrifuged at low-speed (1000 g for 3 min). The supernatant was centrifuged at 400000 g for 30 min. The pellet was resuspended (particulate fraction) and the supernatant recentrifuged at 400000 g. The final supernatant was concentrated with a Centricon-10 microcentrator (soluble fraction). Equivalent amounts of protein were immunoprecipitated with affinity-purified AS antibodies and analysed by SDS/PAGE and fluorography as described in Fig. 3. (*a*) [^{35}S]Methionine-labelled proteins (40 μg of protein). (*b*) [^{3}H]Myristic acid-labelled proteins (40 μg of protein). (From [26].)

Recent studies [24], involving transient expression of α-subunits in COS cells, contradict this idea. Quantitative immunoblotting shows that $\alpha_i 3$, for example, can be expressed at levels of >10-fold that of β-subunits and yet the expressed α-subunits are primarily membrane-associated. It is highly likely that at least certain types of α-subunit are capable of membrane association independent of $\beta\gamma$-subunits. Earlier studies [22] had shown that tryptic cleavage of brain or neutrophil membranes releases α_i and α_o subunits (minus a 1–2 kDa *N*-terminal fragment) from the membranes. Thus α-subunits are, in fact, basically cytosolic proteins anchored to the membrane via their *N*-termini.

Buss *et al.* [25] have found that α_i and α_0, but not α_s, are myristylated. Since this co-translational modification generally involves an *N*-terminal glycine residue, it could explain the anchoring role of the *N*-terminus. To determine the role of myristylation in α-subunit membrane association, we expressed α_s and $\alpha_i 1$ in COS cells, and immunoprecipitated the metabolically labelled products after cell fractionation [26]. Both α-subunits were abundantly expressed and primarily membrane associated, but only $\alpha_i 1$ was found to incorporate [^{3}H]myristic acid (Fig. 3). By site-directed mutagenesis, we altered the second residue of the $\alpha_i 1$ cDNA from glycine to alanine. The mutant protein failed to incorporate [^{3}H]myristate and was found primarily in the soluble fraction (Fig. 4). Nonetheless, the mutant protein could interact with $\beta\gamma$-subunits, since the latter promoted pertussis toxin-catalysed ADP-ribosylation of the mutant α-subunit [26]. These results indicate that myristylation of α_i, but not α_s, subunits is critical for membrane attachment. The basis for membrane attachment of α_s-subunits requires further study.

Both pertussis toxin substrate G-protein α-subunits and G-protein γ-subunits show *C*-terminal sequence similarity with p21 ras proteins [5]. The latter are known to undergo a complex series of *C*-terminal post-translational modifications that are critical for membrane attachment [27,28]. Results with the $\alpha_i 1$ mutant discussed above, make it unlikely that a similar process occurs for α_i-subunits, but the possibility that γ-subunits undergo ras-like *C*-terminal modifications, and that such modifications are important for membrane attachment of $\beta\gamma$-subunits must be considered.

We thank J. Codina, Baylor College of Medicine, for providing G-proteins expressed in *E. coli* and M. Brann, N.I.N.D.S., for collaboration on transient expression of G-proteins.

References

1. Gilman, A. G. (1987) *Annu. Rev. Biochem.* **56**, 615–649
2. Spiegel, A. M. (1987) *Mol. Cell Endocrinol.* **49**, 1–16
3. Birnbaumer, L., VanDongen, A. M., Codina, J., Yatani, A., Mattera, R., Graf, R. & Brown, A. M. (1989) *Soc. Gen. Physiol. Ser.* **44**, 17–54
4. Fong, H. K., Yoshimoto, K. K., Eversole-Cire, P. & Simon, M. I. (1988) *Proc. Natl. Acad. Sci. U.S.A.* **85**, 3066–3070
5. Lochrie, M. A. & Simon, M. I. (1988) *Biochemistry* **27**, 4957–4965
6. Matsuoka, M., Itoh, H., Kozasa, T. & Kaziro, Y. (1988) *Proc. Natl. Acad. Sci. U.S.A.* **85**, 5384–5388
7. Jones, D. T. & Reed, R. R. (1989) *Science* **244**, 790–795
8. Strathmann, M., Wilkie, T. M. & Simon, M. I. (1989) *Proc. Natl. Acad. Sci. U.S.A.* **86**, 7407–7409
9. Bray, P., Carter, A., Simons, C., Guo, V., Puckett, C., Kamholz, J., Spiegel, A. & Nirenberg, M. (1986) *Proc. Natl. Acad. Sci. U.S.A.* **83**, 8893–8897
10. Cerione, R. A., Staniszewski, C., Benovic, J. L., Lefkowitz, R. J., Caron, M. G., Gierschik, P., Somers, R., Spiegel, A. M., Codina, J. & Birnbaumer, L. (1985) *J. Biol. Chem.* **260**, 1493–1500
11. Roof, D. J., Applebury, M. L. & Sternweis, P. C. (1985) *J. Biol. Chem.* **260**, 16242–16249
12. Senogles, S. E., Benovic, J. L., Amlaiky, N., Unson, C., Milligan, G., Vinitsky, R., Spiegel, A. M. & Caron, M. G. (1987) *J. Biol. Chem.* **262**, 4860–4867
13. Goldsmith, P., Rossiter, K., Carter, A., Simonds, W., Unson, C. G., Vinitsky, R. & Spiegel, A. M. (1988) *J. Biol. Chem.* **263**, 6476–6479
14. Goldsmith, P., Gierschik, P., Milligan, G., Unson, C. G., Vinitsky, R., Malech, H. L. & Spiegel, A. M. (1987) *J. Biol. Chem.* **262**, 14683–14688
15. Goldsmith, P., Backlund, P. S., Jr, Rossiter, K., Carter, A., Milligan, G., Unson, C. G. & Spiegel, A. (1988) *Biochemistry* **27**, 7085–7090
16. Sullivan, K. A., Miller, R. T., Masters, S. B., Beiderman, B., Heideman, W. & Bourne, H. R. (1987) *Nature (London)* **330**, 758–760
17. West, R. E., Jr, Moss, J., Vaughan, M., Liu, T. & Liu, T. Y. (1985) *J. Biol. Chem.* **260**, 14428–14430
18. Cerione, R. A., Kroll, S., Rajaram, R., Unson, C., Goldsmith, P. & Spiegel, A. M. (1988) *J. Biol. Chem.* **263**, 9345–9352
19. Simonds, W. F., Goldsmith, P. K., Woodard, C. J., Unson, C. G. & Spiegel, A. M. (1989) *FEBS Lett.* **249**, 189–194
20. Simonds, W. F., Goldsmith, P. K., Codina, J., Unson, C. G. & Spiegel, A. M. (1989) *Proc. Natl. Acad. Sci. U.S.A.* **86**, 7809–7813
21. Yatani, A., Mattera, R., Codina, J., Graf, R., Okabe, K., Padrell, E., Iyengar, R., Brown, A. M. & Birnbaumer, L. (1988) *Nature (London)* **336**, 680–682
22. Eide, B., Gierschik, P., Milligan, G., Mullaney, I., Unson, C., Goldsmith, P. & Spiegel, A. (1987) *Biochem. Biophys. Res. Commun.* **148**, 1398–1405
23. Sternweis, P. C. (1986) *J. Biol. Chem.* **261**, 631–637
24. Simonds, W. F., Collins, R. M., Spiegel, A. M. & Brann, M. R. (1989) *Biochem. Biophys. Res. Commun.* **164**, 46–53
25. Buss, J. E., Mumby, S. M., Casey, P. J., Gilman, A. G. & Sefton, B. M. (1987) *Proc. Natl. Acad. Sci. U.S.A.* **84**, 7493–7497
26. Jones, T. L. Z., Simonds, W. F., Merendino, J. J., Jr, Brann, M. R. & Spiegel, A. M. (1990) *Proc. Natl. Acad. Sci. U.S.A.* **87**, 568–572

27. Hancock, J. F., Magee, A. I., Childs, J. E. & Marshall, C. J. (1989) *Cell* (*Cambridge, Mass*) **57**, 1167–1177
28. Schafer, W. R., Kim, R., Sterne, R., Thorner, J., Kim, S. H. & Rine, J. (1989) *Science* **245**, 379–385

Biochem. Soc. Symp. **56**, 71–80
Printed in Great Britain

G-Proteins and the Inositol Cycle in *Dictyostelium discoideum*

ANTHONY A. BOMINAAR*, JEROEN VAN DER KAAY*, FANJA KESBEKE†, B. EWA SNAAR-JAGALSKA* and PETER J. M. VAN HAASTERT*

**Department of Biochemistry, University of Groningen, Nijenborgh 16, 9747 AG Groningen, The Netherlands and †Cell Biology and Genetics Unit, Zoological Laboratory, Kaiserstraat 63, 2311 GP Leiden, The Netherlands*

Synopsis

The inositol cycle in *Dictyostelium discoideum* was studied both *in vitro* and *in vivo*. The results are compared to the inositol cycle as it is known from higher eukaryotes. Although there is a strong resemblance the cycles are different at some essential points. In comparison to higher eukaryotes, in the cycle in *D. discoideum* the inositol 1,4,5-trisphosphate [Ins(1,4,5)P_3] kinase appears to be absent and there are additional phosphatases which hydrolyse Ins(1,4,5)P_3 via inositol 4,5-bisphosphate [Ins(4,5)P_2] to inositol 4-phosphate (Ins4P). The function of the receptor-stimulated inositol cycle was elucidated using mutants from the *fgd* A complementation group, which are defective in the G-protein α-subunit, responsible for the activation of phosphoinositidase C. These mutants show defects in both chemotaxis and differentiation, suggesting that the stimulation of phosphoinositidase C is the major sensory transduction pathway in *D. discoideum*.

Introduction

Transmembrane signal transduction is characterized to a large extent by the interaction between its components: ligand and receptor at the cell-surface; G-protein subunits at the inner face of the plasma membrane and effector enzymes. The effector enzymes may vary widely depending on the organism and the ligand, and include adenylate cyclase, guanylate cyclase, phosphoinositidase C and ion channels. As a result of these interactions, intracellular second messengers, such as cyclic AMP, cyclic GMP, Ins(1,4,5)P_3, *sn*-1,2-diacylglycerol, Ca^{2+} and K^+ are produced. Besides the interaction between these proteins, to generate second messengers, an extensive interaction between the second messenger systems exists, such that one system modulates another system. The major problem in the understanding of transmembrane signal transduction is the elucidation of the flow of information through this complicated network of interacting molecules.

Transmembrane signal transduction has been studied extensively in the eukaryotic micro-organism *D. discoideum* and appears to be very similar to signal transduction in higher eukaryotes (see Janssens & Van Haastert, 1987). Cyclic AMP is the extracellular signal in *Dictyostelium*, which is comparable with hormones in mammalian cells. Cyclic AMP is detected by surface receptors that have the classical seven putative transmembrane spanning domains of receptors that interact with G-proteins (Klein *et al.*, 1989). The effector enzymes are adenylate cyclase, guanylate cyclase and phosphoinositidase C, and the second messengers interact with target enzymes, such as protein kinases, Ca^{2+} channels and cytoskeletal components. The two main cellular functions of extracellular cyclic AMP in *Dictyostelium* are chemotaxis, to bring the amoeboid cells into a multicellular structure, and cell-type-specific gene expression, to induce cell differentiation in this structure.

In this paper we describe our recent work on the characterization of the inositol cycle in *Dictyostelium*. The metabolism of Ins(1,4,5)P_3 was investigated *in vivo* and *in vitro*, and the stimulation of Ins(1,4,5)P_3 production by receptor and G-protein agonists was demonstrated. The function of the inositol cycle was established mainly by using a mutant which appears to lack a functional G_α-subunit that activates phosphoinositidase C.

Results and Discussion

Metabolism of Ins(1,4,5)P_3 in vitro

In mammalian cells Ins(1,4,5)P_3 is degraded by a 5-phosphatase yielding Ins(1,4)P_2, which is dephosphorylated via Ins4*P* to inositol. A major fraction of Ins(1,4,5)P_3, however, is phosphorylated to inositol 1,3,4,5-tetratrisphosphate [Ins(1,3,4,5)P_4], which is subsequently dephosphorylated by the same 5-phosphatase to Ins(1,3,4)P_3; this InsP_3 isomer is metabolized by a complex pattern of phosphorylations and dephosphorylations (see Berridge & Irvine, 1989). Pilot experiments in *Dictyostelium* suggested that the metabolism of Ins(1,4,5)P_3 could be different from that in mammalian cells.

The dephosphorylation of Ins(1,4,5)P_3 was examined in homogenates of *Dictyostelium* using a mixture of [2-^{3}H]Ins(1,4,5)P_3 and [4,5-^{32}P]Ins(1,4,5)P_3 followed by chromatography of the products on Dowex columns (Van Lookeren Campagne *et al.*, 1988). The ^{3}H radioactivity is associated with the inositol structure, whereas the ^{32}P radioactivity is present predominantly at the 5-position of the ring (85%); thus, detection of ^{3}H radioactivity shows the extent of dephosphorylation, whereas the detection of ^{32}P radioactivity describes the specificity of the dephosphorylation. It was readily observed that the major part of Ins(1,4,5)P_3 dephosphorylation did not occur at the 5-position as in mammalian cells, since essentially all ^{32}P-radioactivity was retained in the InsP_2 product. We concluded that the product was either Ins(1, 5)P_2 or Ins(4,5)P_2. The InsP_2 product was purified, further dephosphorylated by a *Dictyostelium* lysate, and the inositol phosphate product was analysed and identified by h.p.l.c. as Ins4*P*. Thus the major route of Ins(1,4,5)P_3 dephosphorylation in *Dictyostelium* was identified as:

$$\text{Ins}(1,4,5)P_3 \rightarrow \text{Ins}(4,5)P_2 \rightarrow \text{Ins}4P \rightarrow \text{inositol}$$

This route of Ins(1,4,5)P_3 dephosphorylation is present exclusively in the cytosol and dephosphorylates about 80% of the Ins(1,4,5)P_3. The other 20% of Ins(1,4,5)P_3 dephosphorylation is mediated by the mammalian route:

$$\text{Ins}(1,4,5)P_3 \rightarrow \text{Ins}(1,4)P_2 \rightarrow \text{Ins}4P \rightarrow \text{inositol}$$

Between 30 and 50% of the Ins(1,4,5)P_3 5-phosphatase is present in the membrane fraction; all the other phosphatases are present in the cytosol fraction (Van Lookeren Campagne *et al.* 1988).

Since Ins(1,4,5)P_3 and Ins(1,4)P_2 are both dephosphorylated at the 1-position, we investigated the substrate specificity of this enzyme reaction. The same was carried out for the dephosphorylation of Ins(1,4,5)P_3 and Ins(4,5)P_2 at the 5-position (A. A. Bominaar, E. Roovens, J. Van der Kaay & P. J. M. Van Haastert unpublished work). The cytosolic phosphates were partially purified by DEAE–cellulose chromatography, and the dephosphorylation of Ins(1,4,5)P_3, Ins(1,4)P_2, Ins(4,5)P_2, inositol 1-phosphate (Ins1P), and Ins4P was investigated. To obtain a complete picture of the specificity profile, the dephosphorylation of Ins(1,3,4,5)P_4 and Ins(1,3,4)P_3 was also determined, but the products were not identified. The results are summarized in Table 1. At least six enzyme activities involved in the dephosphorylation of Ins(1,4,5)P_3 to inositol can be distinguished. Enzyme 1 is an Ins(1,4,5)P_3 5-phosphatase capable of dephosphorylating Ins(1,3,4,5)P_4, but not Ins(4,5)P_2; the enzyme may be very similar to the mammalian Ins(1,4,5)P_3 5-phosphatase. Enzyme 2 dephosphorylates Ins(1,4,5)P_3 exclusively, at the 1-position. Enzyme 3 is a minor activity and not very stable; preliminary results suggest that it dephosphorylates Ins(1,4,5)P_3 first at the 4-position and subsequently at the 5-position. Other inositol phosphates are not degraded by this column fraction, suggesting that the enzyme is very specific. Enzyme 4 dephosphorylates Ins(1,4)P_2 at the 1-position, with Ins(1,4,5)P_3 also being a substrate, but Ins(1,3,4,5)P_4 and Ins(1,4,5)P_3 are not. The enzyme may be similar to the mammalian inositol polyphosphate 1-phosphatase. The fifth enzyme dephosphorylates Ins(4,5)P_2 at the 5-position. This enzyme is probably Ins(4,5)P_2 specific and does not dephosphorylate Ins(1,4,5)P_3 or Ins(1,3,4,5)P_4; separation from the Ins(1,4,5)P_3 5-phosphatase, however, was not complete. Finally, Ins1P and Ins4P are dephosphorylated by an inositol monophosphate phosphatase; this enzyme is very similar to the mammalian monophosphatase. In summary, *Dictyostelium* probably has at least six phosphatases that participate in the dephosphorylation of Ins(1,4,5)P_3. Three enzymes appear to be similar to their mammalian counterparts: Ins(1,4,5)P_3 5-phosphatase, Ins(1,4)P_2 1-phosphatase and Ins$P$ phosphatase. The other three may be unique for *Dictyostelium*: the Ins(1,4,5)P_3 1-phosphatase, the Ins(1,4,5)P_3 4 → 5-bisphosphatase and the Ins(4,5)P_2 5-phosphatase. It is likely that this complex dephosphorylation pattern is more than just degradation. InsP_2 isomers may have signal transducing functions. Furthermore, the trifurcation in the dephosphorylation of Ins(1,4,5)P_3 opens ways for fine-regulation, and the system could be compartmentalized or under developmental control.

In contrast to the complex dephosphorylation of Ins(1,4,5)P_3, the phosphorylation of Ins(1,4,5)P_3 is extremely simple: all experimental evidence indi-

Table 1. *Enzymology of Ins(1,4,5)P_3 dephosphorylation*

The high-speed supernatant of a *Dictyostelium* lysate was chromatographed on a DEAE–cellulose column that was eluted with a linear gradient of NaCl. The fractions were assayed for the dephosphorylation of Ins(1,3,4,5)P_4, Ins(1,4,5)P_3, Ins(1,3,4)P_3, Ins(4,5)P_2, Ins(1,4)P_2, Ins1*P*, Ins3*P*, and Ins4*P*; the products of all reactions were identified [with the exception of the reactions with Ins(1,3,4,5)P_4 and Ins(1,3,4)P_3].

Enzyme					
No.	Name	Elution	Reaction	Co-substrate	No substrate
1	Ins(1,4,5)P_3 5-phosphatase	0.1 M	1,4,5 → 1,4	1,3,4,5	4,5
2	Ins(1,4,5)P_3 1-phosphatase	Wash	1,4,5 → 4,5		1,3,4,5; 1,3,4; 1,4
3	Ins(1,4,5)P_3 4 → 5-phosphatase	0.3 M	1,4,5 → 1	—	1,3,4,5; 1,4; 1,3,4
4	Ins(1,4)P_2 1-phosphatase	0.1 M	1,4 → 4	1,3,4	1,3,4,5; 1,4,5
5	Ins(4,5)P_2 5-phosphatase	0.1 M	4,5 → 4	—	1,3,4,5; 1,4,5
6	Ins1*P* phosphatase	0.1 M	$\text{Ins}_x P \rightarrow \text{Ins}$	$x \neq 2$	$x = 2$

cates that the appropriate kinases are absent. The Ins(1,4,5)P_3 3-kinase could not be detected under conditions where the rat brain enzyme was very active and Ins(1,3,4,5)P_4 and Ins(1,3,4)P_3 were not detectable in extracts from [^{3}H]inositol-labelled cells (Van Haastert *et al.*, 1989).

Metabolism of [^{3}H]inositol in vivo

[^{3}H]Inositol was introduced into *Dictyostelium* cells by electroporation. The protocol was optimized to generate small holes, just large enough for inositol to pass, but impermeant to charged molecules such as Ins(1,4,5)P_3 or ATP (Van Haastert *et al.*, 1989). This method of labelling cells with [^{3}H]inositol is very efficient (2.5% within 10 min) compared with metabolic labelling (0.2% in 6 h). The metabolism of [^{3}H]inositol was then followed by the analysis of the inositol phospholipids by thin-layer chromatography and the analysis of the water-soluble compounds by h.p.l.c. (J. Van der Kaay, R. Draifer & P. J. M. Van Haastert unpublished work). Two h.p.l.c. systems were used: Ins(1,4,5) P_3 levels were analysed by reverse phase ion-pair chromatography and the other inositol phosphates by ion-exchange chromatography. A significant amount of the radioactivity is incorporated into unidentified compound(s) that elute from Dowex columns in the InsP_3/InsP_4 fraction, from reverse-phase columns in the inositol hexakisphosphate (InsP_6) fraction and that elute only partially from the h.p.l.c. SAX column (Van Haastert *et al.*, 1989). It should be mentioned that we have not yet identified the inositol phospholipids in terms of polar headgroup and fatty acid composition. This might be relevant, since *Dictyostelium* seems to lack arachidonic acid (MacDonald & Weeks, 1985).

The results (Fig. 1) demonstrate that after pulse-labelling with [^{3}H]inositol radioactivity was rapidly incorporated into phosphatidylinositol (PtdIns); a

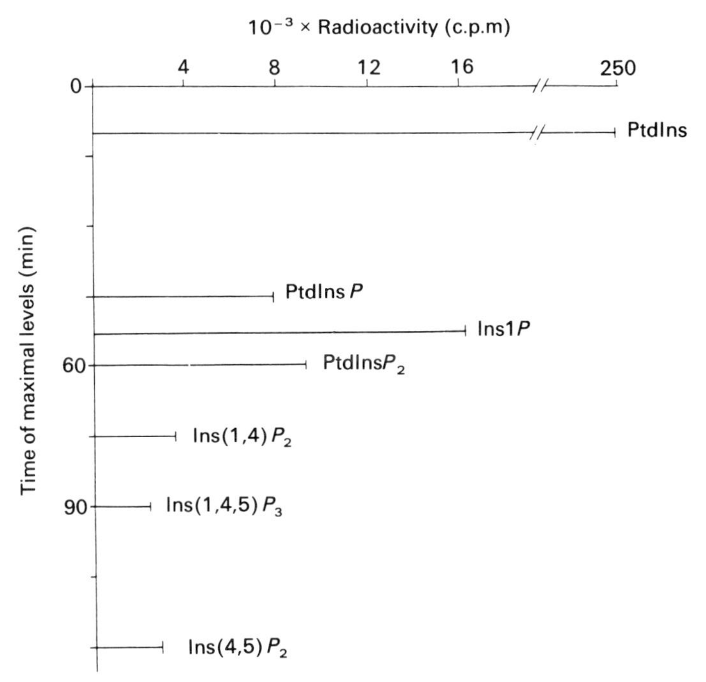

Fig. 1. *Dynamics of the inositol cycle*

Cells were pulse-labelled with [^{3}H]inositol by electroporation. Samples were withdrawn at 15 min intervals and analysed for intracellular levels of inositol-containing compounds. Since there is considerable secretion of [^{3}H]inositol-containing compounds, the radioactivity in all compounds reaches a maximum. The Figure presents the magnitude and time at which the maximal radioactivity in each compound was reached.

maximum was reached in 10 min followed by a decline to half-peak levels after about 60 min. Radioactivity was subsequently found in the phospholipids PtdIns$P$ and PtdInsP_2 with maxima at 45 and 60 min, respectively, after pulse-labelling. These kinetics are in perfect agreement with the lipid part of the inositol cycle where PtdIns is formed from Ins and CDP-DG, and PtdIns$P$ and PtdInsP_2 from the phosphorylation of PtdIns. The first water-soluble inositol phosphate co-chromatographed with Ins1P; the radioactivity was maximal at 50 min after pulse-labelling, well before Ins(1,4)P_2 was formed. This suggests that there is a phosphoinositidase C activity that acts on PtdIns. The second water-soluble product co-chromatographed with Ins(1,4)P_2 and reached a maximum at 75 min. Ins(1,4)P_2 was formed well before Ins(1,4,5)P_3, suggesting that it was not formed through the dephosphorylation of Ins(1,4,5)P_3, but by a phosphoinositidase C reaction acting on PtdIns$P$. [^{3}H]Ins(1,4,5)P_3 was, with a maximum at 90 min, the next water-soluble radioactive compound that could be detected, suggesting that it was formed from PtdInsP_2. The last compound was Ins(4,5)P_2, which is in accordance with its proposed Ins(1,4,5)P_3 source. The kinetics of Ins4P formation are presently unknown, because it was not possible to obtain a complete separation of the small amount of Ins4P from the bulk of Ins1P.

The results of the experiments on the metabolism of Ins(1,4,5)P_3 *in vitro* and inositol *in vivo* provide a nearly complete picture of the inositol cycle in *Dictyostelium*, which is summarized in Fig. 2.

Stimulation of Ins(1,4,5)P_3 production by cyclic AMP and guanosine 5'-[γ-thio]triphosphate (GTP[S])

Cells were labelled with [^{3}H]inositol as described above, incubated for 45 min to allow for the incorporation of [^{3}H]inositol into PtdInsP_2, permeabilized with saponin (Europe-Finner & Newell, 1986), and stimulated with either the receptor agonist cyclic AMP or the G-protein agonist GTP[S]. Both agonists induce a rapid increase of [^{3}H]Ins(1,4,5)P_3 to about 145% of basal levels, with a maximum at about 6 s after stimulation. Basal levels were recovered at 20–30 s after stimulation (Van Haastert *et al.*, 1989). This response could be measured in unlabelled cells as well, when Ins(1,4,5)P_3 was extracted and the amount determined by an isotope dilution assay (Van Haastert, 1989). The results were similar (Fig. 3), but absolute Ins(1,4,5)P_3 levels show that the mean intracellular concentration increased from 3.3 μM to a maximum of 5.5 μM. Thus, the relative increase is small, but taking into account the submicromolar concentrations of Ins(1,4,5)P_3 that induce Ca^{2+} release, the absolute increase is considerable.

Since the pool size of Ins(1,4,5)P_3 is about ten times smaller than the pool of PtdInsP_2, it is not surprising that we have never been able to detect a receptor-stimulated decrease of PtdInsP_2. Unfortunately, this implies that we have no complete evidence that the increase of Ins(1,4,5)P_3 is the result of a receptor- and G-protein-stimulated phosphoinositidase C activity.

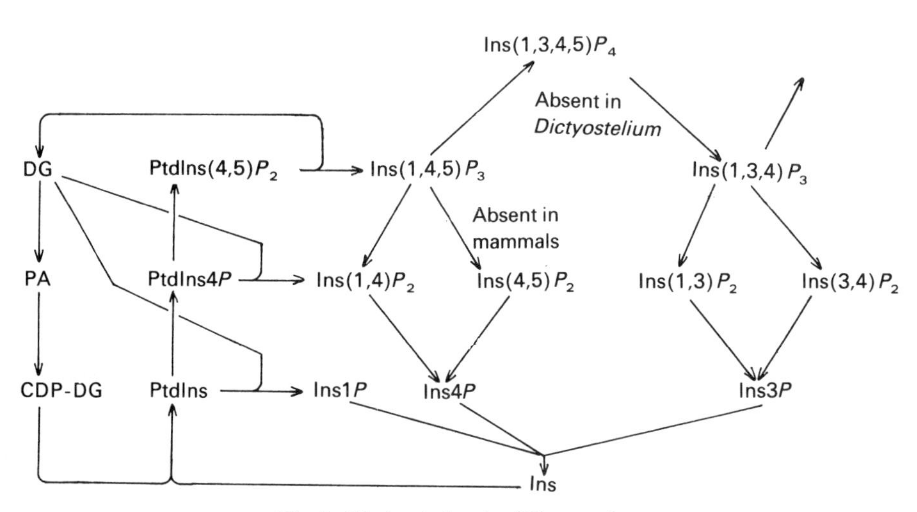

Fig. 2. *The inositol cycle of Dictyostelium*

A small portion of Ins(1,4,5)P_3 may also be dephosphorylated to Ins1P via Ins(1,5)P_2. Abbreviations: DG, diacylglycerol; PA, phosphatidic acid.

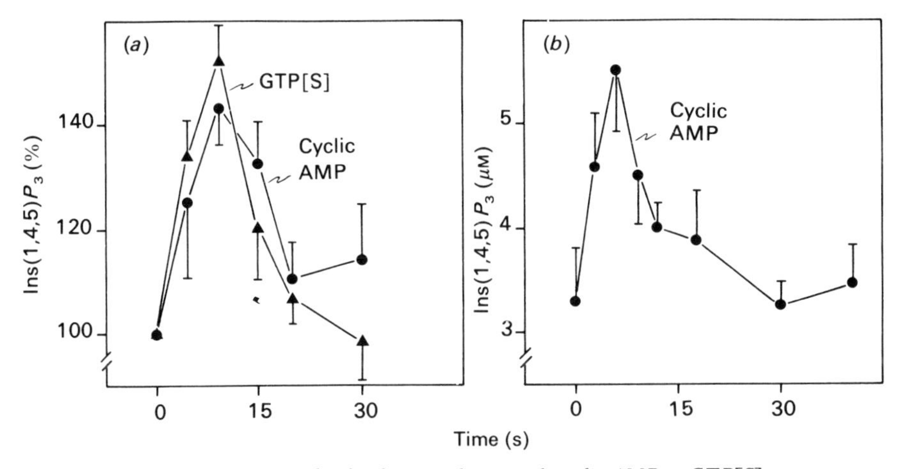

Fig. 3. *Ins(1,4,5)P_3 levels after stimulation with cyclic AMP or GTP[S]*

(*a*) Relative levels of [^{3}H]Ins(1,4,5)P_3 after stimulation of saponin-permeabilized cells with cyclic AMP or GTP[S]. (*b*) Absolute Ins(1,4,5)P_3 levels of cells after stimulation with cyclic AMP.

Function of the inositol cycle in Dictyostelium

Several mutants of *Dictyostelium* were analysed for the receptor and G-protein stimulation of Ins(1,4,5)P_3 production (P. J. M. Van Haastert, unpublished work). The biochemical and genetic characterization of mutants of the complementation group *fgd* A (Coukell *et al.*, 1983) are presented in Fig. 4. Biochemical data suggested a defect in the interaction between receptor and G-protein (Kesbeke *et al.*, 1988; Snaar-Jagalska *et al.*, 1988). Genetic data imply that this defect is the result of alterations in the gene encoding the G-protein $G_\alpha 2$. Strain HC-85 has a 2.2 kb deletion in the coding region of the gene (Kumagai *et al.*, 1989). Mutants of the *fgd* A complementation group are defective in both cyclic AMP and GTP[S] stimulation of Ins(1,4,5)P_3 levels (Table 2). These results strongly suggest that the defective G-protein is involved in the receptor-mediated activation of phosphoinositidase C. This hypothesis should be confirmed by experiments which measure a GTP stimulation of phosphoinositidase C activity *in vitro*.

Assuming that the hypothesis is correct, the phenotype of the mutant provides the function of the receptor-stimulated inositol cycle in *Dictyostelium*.

Table 2. *Cyclic AMP and GTP[S] stimulation of Ins(1,4,5)P_3 levels in the mutant fgd A*

Wild-type and mutant cells were labelled with [^{3}H]inositol, permeabilized with saponin and stimulated with cyclic AMP or GTP[S]. [^{3}H]Ins(1,4,5)P_3 levels were determined by h.p.l.c.

Stimulus	Ins(1,4,5)P_3 levels (% of control)	
	Wild-type	*fgd* A
Cyclic AMP	146 ± 11*	101 ± 4 NS
GTP[S]	148 ± 13*	102 ± 8 NS

* Significantly different from basal levels at $P < 0.01$; $n = 4$; NS, not significant.

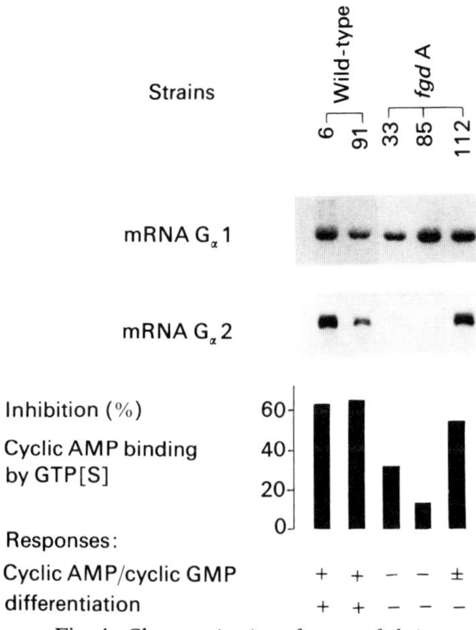

Fig. 4. *Characterization of mutant fgd A*

Three strains of the *fgd* A complementation group and the two corresponding parent strains were starved for 4 h. The top panel shows a Northern blot of mRNA probed with $G_\alpha 1$ cDNA and $G_\alpha 2$ cDNA; using the same batch of cells, the inhibition of cyclic AMP binding to membranes by GTP[S], the cyclic AMP-mediated activation of adenylate and guanylate cyclase, and the differentiation were measured (bottom panel). The Northern blots were probed by M. Pupillo (Baltimore).

The characteristics of mutant *fgd* A are summarized in Table 3. Functional cyclic AMP receptors are present in terms of binding, covalent modification and down-regulation. However, none of the second messenger responses is induced in the mutant, and cyclic AMP does not induce chemotaxis or differentiation. Thus, there is a complete blockade of all signal transduction in mutant *fgd* A. It was, therefore, surprising to find that GTP stimulation of adenylate cyclase was still normal in membranes from the mutant (Kesbeke *et al.*, 1988). This suggests that the defective G-protein is not directly involved in the regulation of adenylate cyclase. This observation leads to two conclusions:

Table 3. *Characterization of mutant fgd A*

Characteristics of *fgd* A
Surface cyclic AMP receptors are present
Normal cyclic AMP-induced receptor phosphorylation
Normal cyclic AMP-induced receptor down-regulation
No cyclic AMP-induced cyclic AMP formation
No cyclic AMP-induced cyclic GMP formation
No cyclic AMP-induced chemotaxis
No cyclic AMP-induced differentiation
Normal GTP[S] stimulation of adenylate cyclase
No GTP[S] stimulation of Ins(1,4,5)P_3 production

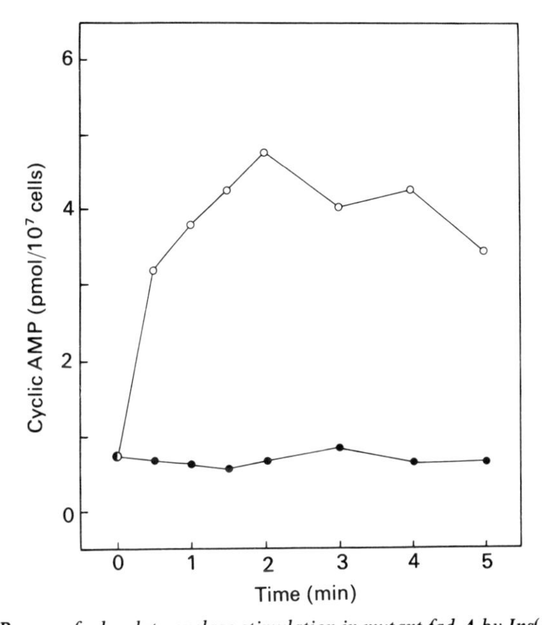

Fig. 5. *Rescue of adenylate cyclase stimulation in mutant fgd A by Ins(1,4,5)P_3*
Saponin-permeabilized mutant cells were stimulated with the receptor agonist 2′-dibutyryl in the presence (○) or absence (●) of Ins(1,4,5)P_3 and the accumulation of cyclic AMP was measured.

there must be another G-protein and the G-protein that stimulates phospholipase C is indirectly involved in the regulation of adenylate cyclase.

This last hypothesis was tested by the experiment shown in Fig. 5. Mutant *fgd* A cells were permeabilized with saponin and stimulated with cyclic AMP in the absence or presence of Ins(1,4,5)P_3 and the production of cellular cyclic AMP levels was measured. Cyclic AMP or Ins(1,4,5)P_3 alone did not alter cellular cyclic AMP levels, but a strong increase was observed when both compounds were present. Subsequent experiments suggest that cyclic AMP acts on the surface receptors, whereas Ins(1,4,5)P_3 is required intracellularly.

Conclusions

The micro-organism *Dictyostelium* has an inositol cycle that is reminiscent of the inositol cycle of mammalian cells. The main differences are found in the metabolism of Ins(1,4,5)P_3: (i) the phosphorylation to Ins(1,3,4,5)P_4 seems to be absent (and thus the complex metabolism of this compound); and (ii) the dephosphorylation of Ins(1,4,5)P_3 is more complex and involves three additional enzyme activities that have not been described in mammalian cells. The inositol cycle is probably the major signal transduction pathway in *Dictyostelium* as was demonstrated with mutant *fgd* A which apparently lacks the functional G-protein required to activate phosphoinositidase C.

We have not yet completely identified the inositol cycle of *Dictyostelium.* During the analysis of the water-soluble compounds, large quantities of radioactivity that could not be identified were detected; these compounds may contain up to 50% of the water-soluble radioactivity. Furthermore, it is still largely unknown how InsP_6 is formed in *Dictyostelium*. Finally, identification of some compounds, such as the fatty acid and polar headgroup of the inositol phospholipids, is necessary. Whereas the identification of the inositol cycle may be well established, the identification and characterization of the participating enzymes are still in their infancy. The enzyme phosphoinositidase C has not been detected in assays *in vitro* (Irvine *et al.*, 1980). Finally, the precise function of Ins(1,4,5)P_3 in the cell is not completely known; it does release Ca^{2+} from non-mitochondrial stores (Europe-Finner & Newell, 1986), but other functions cannot be excluded. The latter is suggested by the complex dephosphorylation of Ins(1,4,5)P_3. These details of the inositol cycle might be relevant since the inositol cycle controls signal transduction, chemotaxis and differentiation in *Dicytostelium*, as is shown by mutant *fgd* A.

This study was supported by grants of the N.W.O. C. and C. Huygens Fund and the N.W.O. council for medical research.

References

Berridge, M. J. & Irvine, R. F. (1989) *Nature (London)* **341**, 197–205
Coukell, M. B., Lappano, S. & Cameron, A. M. (1983) *Dev. Genet.* **3**, 283–297
Europe-Finner, G. N. & Newell, P. C. (1986) *Biochim. Biophys. Acta* **887**, 335–340
Irvine, R. F., Letcher, A. J., Brophy, P. J. & North, M. J. (1980) *J. Gen. Microbiol.* **121**, 487–495
Janssens, P. M. W. & Van Haastert, P. J. M. (1987) *Microbiol. Rev.* **51**, 396–418
Kesbeke, F., Snaar-Jagalska, B. E. & Van Haastert, P. J. M. (1988) *J. Cell. Biol.* **107**, 521–528
Klein, P., Sun, T. J., Saxe, C. L., Kimmel, A. R. & Devreotes, P. N. (1989) *Science* **241** 1467–1472
Kumagai, A., Pupillo, M., Gunderson, R., Mike-Lye, R., Devreotes, P. N. & Firtel, R. A. (1989) *Cell (Cambridge, Mass.)* **57**, 265–275
MacDonald, J. I. S. & Weeks, G. (1985) *Biochim. Biophys. Acta* **834**, 301–307
Snaar-Jagaska, B. E., Kesbeke, F., Pupillo, M. & Van Haastert, P. J. M. (1988) *Biochem. Biophys. Res. Commun.* **156**, 757–761
Van Haastert, P. J. M. (1989) *Anal. Biochem.* **177**, 115–119
Van Haastert, P. J. M., De Vries, M. J., Penning, L. C., Roovers, E., Van der Kaay, J., Erneux, C. & Van Lookeren Campagne, M. M. (1989) *Biochem. J.* **258**, 577–586
Van Lookeren Campagne, M. M., Erneux, C., Van Eijk, R. & Van Haastert, P. J. M. (1988) *Biochem. J.* **254**, 343–350

Biochem. Soc. Symp. **56**, 81–84
Printed in Great Britain

A Likely History of G-Protein Genes

JAMES B. HURLEY

Howard Hughes Medical Institute, SL-15, University of Washington, Seattle, WA 98175, U.S.A.

Introduction

A plethora of cellular responses are mediated by G-proteins. In fact, the recent recognition that G-protein genes make up an extremely large and diverse family in itself suggests such a wide range of physiological applications. G-proteins even have multiple functions. For example, a G_i α-subunit can both inhibit adenylate cyclase and activate K^+ channels (Freissmuth *et al.*, 1989). This paper describes a likely model for how G-protein genes evolved to perform multiple and various functions. The model is based on estimates of the average number of third-base substitutions that must have occurred during the time that two genes have been separate.

Results

Coding regions of all available G-protein gene sequences were aligned using the NUCALN program (Wilbur & Lipman, 1983) and the number of transitions and transversions at third-base positions of corresponding codons determined manually. These data were used to estimate the average number of third-base substitutions that occurred at third-base positions from the time that two G-protein genes duplicated and diverged. Estimates were performed by the method of Kimura (1981) and the results of the analyses of human gene sequences are shown in Fig. 1.

Alignment of G-protein amino acid sequences and DNA sequences can be performed with reasonable certainty because the proteins have been highly conserved throughout evolution. For example, the *Drosophila* G_i and G_o α-subunits and the *Drosophila* β-subunit have exactly the same number of amino acids as their mammalian counterparts (Provost *et al.*, 1988; Yarfitz *et al.*, 1988; de Sousa *et al.*, 1989). In fact, the *Drosophila* and mammalian G-protein amino acid sequences are ~80% similar to each other. The G_s α-subunit could only be aligned with other α-subunits with considerably less certainty, so it was excluded from this analysis.

Fig. 1 shows that the two genes most closely related to each other by this analysis are rod transducin and $\alpha_i 2$. α_z and α_0 are slightly less related to this group. Surprisingly, the genes that became rod and cone transducins appear to have diverged from each other long ago. It is difficult to assign to this type of analysis a time scale. However, as a rough estimate, an average of ~0.8 substitutions per third-base position separate mammalian α- and β-subunits from

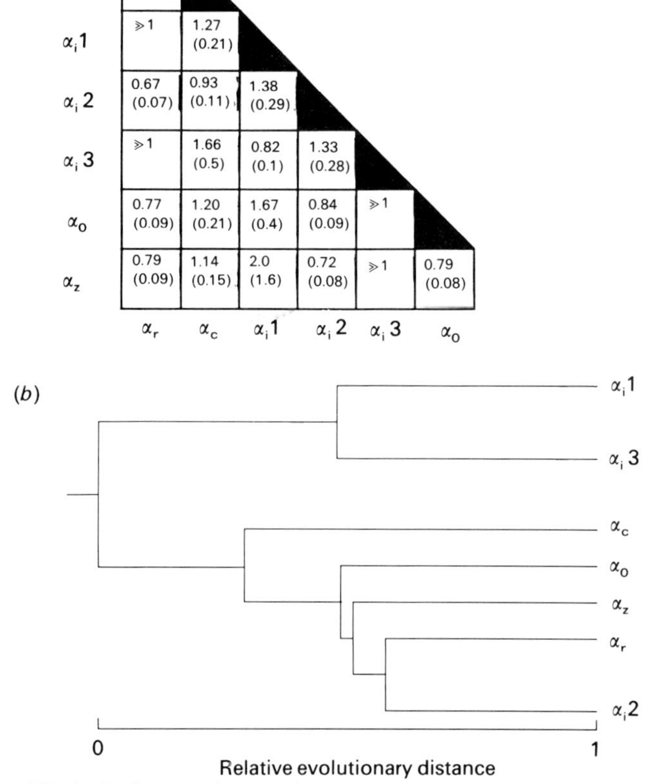

Fig. 1. *Evolutionary history of G-proteins based on third-base substitutions*

(*a*) Estimates of the average number of third-base substitutions between pairs of human G-protein α-subunit nucleotide sequences as calculated by the method of Kimura (1981). References to the sequences used in this analysis are found in Lochrie & Simon (1988). (*b*) A likely evolutionary history of G-protein genes based on the values shown in (*a*). The 'relative evolutionary distance' is assumed to be proportional to the average number of third-base substitutions.

the corresponding *Drosophila* genes. Thus it is likely that the rod and cone transducin α-subunit genes diverged from each other before vertebrates diverged from invertebrates, about 500 million years ago.

The most surprising result from this analysis is the relationship between the three different forms of G_i α-subunits. Although all three have amino acid sequences ~90% similar to each other, their genes appear to have diverged long ago. In fact, the $G_i 2\alpha$ gene seems only distantly related to the $G_i 1\alpha$ and $G_i 3\alpha$ genes. Even though these genes appear to have diverged long ago, G_i α-subunits have multiple functions that have been conserved. For example, they interact with the same receptors, β- and γ-subunits, ion channels and, by some mechanism, adenylate cyclase (Freissmuth *et al.*, 1989). There are clearly multiple constraints on the amino acid sequences encoded by these genes.

The results presented here must be interpreted with caution. In order that the number of third-base substitutions be proportional to the time over which genes have been separated two assumptions must be valid. (i) Mutations in these genes must have occurred independently. Gene conversion events may have occurred, but there is no way to directly determine this. If these types of events did occur, the genes being compared would appear to have duplicated and diverged more recently than they actually did. During the third-base substitution analysis, no 'patches' of sequence identity in the genes being compared were found, suggesting that neither partial gene conversion events nor exon-exchange events occurred. (ii) Mutations must have accumulated at the same rate in all the genes being analysed. Since G-protein genes are present on many different chromosomes (Lochrie & Simon, 1988), they may not be subject to the same mutagenic forces or constraints.

The results of this analysis present a picture strikingly different from the one generated by a similar type of analysis of amino acid sequences (Itoh *et al.*, 1988). From the knowledge of the mechanisms of molecular evolution available at this time, it does not appear possible to decide which scheme more accurately represents G-protein gene evolution. Clearly, there are functional constraints on protein structure that very likely distort the analysis and interpretation of amino acid sequence comparisons. However, the precautions listed above for third-base substitution analyses are equally significant.

Regardless of the exact order in which G-protein genes appeared during evolution, it is clear that many G-protein genes have been adapted for very specific functions. For example, a specific G_s α-subunit gene is expressed only in olfactory tissue (Jones & Reed, 1989). Different transducin α-subunits are found either in rod cells or in cone cells of the retina (Lerea *et al.*, 1986). This level of cell type specificity is not unprecedented, particularly in sensory cells. Each type of cone cell in the human retina expresses a different opsin gene which imparts on the cell a unique spectral sensitivity. In fact, blue cone cells even produce a form of phosphodiesterase inhibitor subunit different from that of red and green cone cells (Hamilton & Hurley, 1990). Rod and cone functions are basically the same, but they have different spectral sensitivities, different overall sensitivity and different physiological kinetics. The molecular basis of these differences has not been determined. Rods and cones express different opsin genes, different transducin genes, different phosphodiesterase catalytic subunit genes and different phosphodiesterase inhibitor subunit genes. Therefore, biochemical differences between any of these components of the cyclic GMP cascade in rod and cone photoreceptors could account for the dramatic physiological differences between these retinal cell types.

Other G-proteins also appear to be expressed differentially. In *Drosphila*, the G_o α-subunit is found throughout the nervous system. The G_i α-subunit is found in the nuclei of all cells, but it is most abundant in nurse cell nuclei within the ovary (N. Provost, unpublished work). Even within the α_i-subunit family in mammals, there is tissue specificity (Jones & Reed, 1987). The biochemical contributions of these different α-subunits to the physiological differences between different cells and tissues is unknown.

The ways that cells use G-proteins with subtle differences in activity or function are certain to be of physiological significance, because G-protein genes that have been separated for long periods of evolutionary history have been well preserved. Identification of those subtle differences and their significance will require the combined efforts of biochemists and physiologists.

References

de Sousa, S., Hoveland, L. L., Yarfitz, S. & Hurley, J. B. (1989) *J. Biol. Chem.* **264**, 18544–18551
Freissmuth, M., Casey, P. J. & Gilman, A. G. (1989) *FASEB J.* **3**, 2125–2131
Hamilton & Hurley, J. B. (1990) *J. Biol. Chem.* in the press
Itoh, H., Toyama, R., Kozasa, T., Tsukamoto, T., Matsuoka, M. & Kaziro, Y. (1988) *J. Biol. Chem.* **263**, 6656–6664
Jones, D. T. & Reed, R. R. (1987) *J. Biol. Chem.* **262**, 14241–14249
Jones, D. T. & Reed, R. R. (1989) *Science* **19**, 790–795
Kimura, O. (1981) *Proc. Natl. Acad. Sci. U.S.A.* **78**, 454
Lerea, C. L., Somers, D. E., Hurley, J. B., Klock, I. B. & Bunt-Milam, A. H. (1986) *Science* **234**, 77–80
Lochrie, M. & Simon, M. I. (1988) *Biochemistry* **27**, 4957–4965
Provost, N. M., Somers, D. E. & Hurley, J. B. (1988) *J. Biol. Chem.* **263**, 12070–12076
Wilbur, W. J. & Lipman, D. J. (1983) *Proc. Natl. Acad. Sci. U.S.A.* **80**, 726–730
Yarfitz, S., Provost, N. M. & Hurley, J. B. (1988) *Prov. Natl. Acad. Sci. U.S.A.* **85**, 7134–7138

Biochem. Soc. Symp. **56**, 85–101
Printed in Great Britain

The Role and Mechanism of the GTP-Binding Protein G_E in the Control of Regulated Exocytosis

BASTIEN D. GOMPERTS, YASMIN CHURCHER, ANNA KOFFER, IJSBRAND M. KRAMER, TOM LILLIE and PETER E. R. TATHAM

Department of Physiology, University College London, University Street, London WC1E 6JJ, U.K.

Synopsis

Recent advances in the understanding of the terminal stages of the pathway of regulated secretion (exocytosis) have depended in large part on the use of permeabilized secretory cells in which the cytosol is directly accessible to manipulation from the outside. As well as Ca^{2+}, a role for GTP, and hence a GTP-binding protein (G_E), is plainly apparent in the exocytotic mechanism of some cells. In metabolically depleted and permeabilized mast cells, the combination of Ca^{2+} and a guanine nucleotide are a sufficient stimulus for the release of the total cell content of histamine and hexosaminidase. ATP is not required and so a phosphorylation reaction does not comprise a necessary step in the terminal pathway. Phosphorylating nucleotides, however, can modulate the exocytotic reaction in a manner that suggests that the reverse reaction, namely protein dephosphorylation, might be an enabling step. Support for this idea is derived from experiments in which the cells are conditioned to become responsive to Ca^{2+} alone; in these circumstances the dependence of secretion on the presence of a guanine nucleotide can be reimposed by okadaic acid, an inhibitor of protein phosphatases.

Introduction

Intracellular effects of exocytosis

It may seem a bit odd to begin an essay on the role of GTP-binding proteins in secretion with a discussion of Ca^{2+}. This is because Ca^{2+} has for so long been recognized as the key player in this area (Douglas, 1974) and the reason that it appeared to hold the centre stage is that it is simply the most evident and the most easily manipulated of all the second messengers. This was so even before the recent developments which have allowed for direct measurement of its intracellular concentration with fluorescent dyes (Tsien, 1986; Cheek, 1989) and the precise regulation of its concentration in the cytosol of permeabilized cells (Gomperts & Fernandez, 1985). Nevertheless, it was already evident that there are other second messenger systems which can activate secretion in some specific situations. Thus, in 1974 it was shown that amylase secretion from rat parotid glands can be stimulated with β-adrenergic ligands and is associated with an elevation of cyclic AMP (Dormer & Ashcroft, 1974; Leslie *et al.*, 1976) [and latterly shown to occur at resting levels of intracellular Ca^{2+} (McMillian *et al.*, 1988)]. Secretion of parathyroid hormone has long been associated with a reduction in the concentration of extracellular Ca^{2+} (Sherwood *et al.*, 1966; Brown *et al.*, 1987) and this has since been shown to be linked also to a reduction in the level of intracellular Ca^{2+} (Nemeth & Scarpa, 1986; Shoback *et al.*, 1983). This is not a unique oddity since secretion of some other polypeptide hormones such as human placental lactogen (Choy & Watkins, 1976; Handwerger *et al.*, 1981) and renin (Park *et al.*, 1986; Fray *et al.*, 1987) also correlates with a decrease, not an increase, in the cytosolic concentration of Ca^{2+}.

All these examples concern the stimulation of secretion in a situation in which the level of a soluble second messenger is altered as a result of stimulation. It was only with the development of techniques for selective plasma membrane permeabilization that one could begin to ask questions about possible effector mechanisms which might operate primarily, or at least in part, by regulation of the affinity of an intracellular receptor (such as a GTP-binding protein) for a soluble ligand whose concentration remains more or less static.

G_E, a GTP-Binding Protein Mediating Exocytosis

It is by now quite clear that for some (but probably not all) secretory cells there is a regulatory role for a GTP-binding protein (G_E) late in the pathway leading to fusion of secretory vesicles with the plasma membrane (Gomperts, 1989; Gomperts, 1990). As nearly all the evidence for involvement of G_E comes from experiments with permeabilized cells, it is pertinent to point out that intact RBL-2H3 cells treated with agents which cause a decrease in GTP levels, fail to undergo secretion either in response to receptor-directed ligands (specific antigen to cross-link cell-surface IgE) or to the Ca^{2+} ionophore A23187 (Marquardt *et al.*, 1987; Wilson *et al.*, 1988). These agents (such as mycophenolic acid or ribavirin) cause the selective depletion of cytosol levels of GTP as a result of their propensity to inhibit IMP dehydrogenase (Johnson & Mukku, 1979). While such treatment can readily be understood to inhibit G_P-linked receptor-mediated processes involving phospholipase C, the indications are that the Ca^{2+}-induced secretory process also requires the presence of GTP. A similar argument was used earlier as evidence for a role for GTP in the stimulation of adenylyl cyclase in intact cells (Johnson & Mukku, 1979).

Accessing the Cytosol

The manner in which the activity of G_E is expressed varies among cell types (see Fig. 1). While such variation might be real, it is also now quite clear that the many different procedures for permeabilization allow different extents of leakage and thus, in themselves, introduce variations in cell responsiveness (Gomperts & Fernandez, 1985; Koffer & Gomperts 1989). For example, it matters greatly whether the cells have been permeabilized to the extent of allowing selective dialysis of the interior, permitting exchange of ions and low

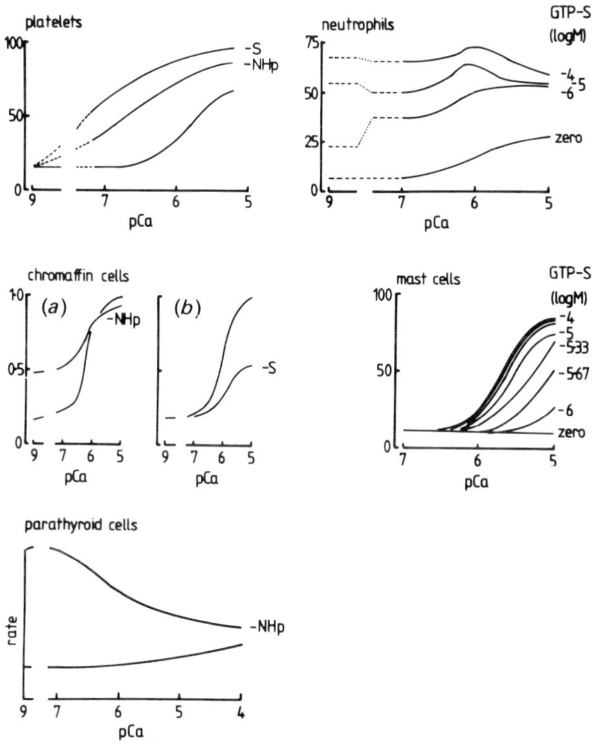

Fig. 1. *Interaction between Ca^{2+} and guanine nucleotides in the exocytotic process of a number of different secretory cells*

The graphs indicate the dependence of secretion on [Ca^{2+}] from various secretory cells. [Ca^{2+}] was regulated by the use of Ca^{2+}/EGTA buffers; except for the mast cells, ATP was also provided. Platelets [permeabilization by high voltage discharge (HVD)] (Haslam & Davidson, 1984; Knight & Scrutton, 1986): effect of 100 μM-guanosine 5′-[$\beta\gamma$-imido]triphosphate (Gpp[NH]p) (−NHp) and 10 μM-GTP[S] (−S). Chromaffin cells (*a*): (digitonin permeabilization), effect of 100 μM-Gpp[NH]p (−NHp) (Bittner *et al.*, 1986); (*b*): (HVD), effect of 80 μM-GTP[S] (−S) (NB Gpp[NH]p (80 μM) was without effect in this experiment) (Knight & Baker, 1985). Parathyroid cells (HVD): effect of 10 μM-Gpp[NH]p (−NHp) (Oetting *et al.*, 1987). Neutrophils (Sendai virus permeabilization) (Barrowman *et al.*, 1987). Mast cells (streptolysin O permeabilization) (Howell *et al.*, 1987). From Gomperts (1990).

molecular mass solutes, or whether the permeabilization allows complete washout of all soluble components (see Table 1).

In the case of mast cells, we have found that much lower concentrations of activating effectors are capable of inducing exocytosis after permeabilization with streptolysin O than with ATP^{4-} (Fig. 2*a* and *b*) (Koffer & Gomperts, 1989). These results suggest the existence of an inhibitory cytosolic factor which is rapidly released from streptolysin O-permeabilized cells, but which is retained in cells permeabilized by ATP^{4-}, and is therefore present also in intact cells. Since agonist-induced exocytosis from intact mast cells and related RBL-2H3 cells occurs at and even below pCa 6.5 (Beaven *et al.*, 1987; Beaven & Cunha-Melo, 1988), the interaction between this inhibitory factor and the Ca^{2+} or GTP-binding protein is likely to be under control of the receptor process.

Table 1. *Methods of cell permeabilization*

The filtration dimensions are given only as a very rough guide. For any reagent or method there will be wide variation depending on membrane composition and other conditions. Furthermore, the different methods used to assess the filtration properties of membrane lesions must necessarily give different results. Abbreviation: LDH, lactate dehydrogenase.

	Effective filtration diameter	Method of assessment	Reference
Methods permitting selective dialysis of ions and low molecular mass solutes			
Sendai virus	Approx. 1 nm	Exclusion of fluorescent peptides	Impraim *et al.* (1980)
Staphylococcal α-toxin	2–3 nm	Electron microscopy	Füssle *et al.* (1981)
High-voltage discharge	2–4 nm	Rate of uptake and efflux of various markers in chromaffin cells	Knight & Baker (1982)
ATP^{4-} (low concentrations)	Variable dimensions (α-[ATP^{4-}])	Rate of ^{32}P-metabolite efflux and ^{57}Co, HEDTA uptake, membrane conductance	Cockcroft & Gomperts (1979), Bennett *et al.* (1981) Tatham & Lindau (1990)
Methods permitting exchange of macromolecules			
ATP^{4-} (high concentrations)	Variable dimensions	Efflux of actin	Koffer & Gomperts (1989)
Lysolecithin	Variable dimensions	RNAase (M_r 14 000) uptake and LDH efflux	Miller *et al.* (1978)
Plant glycosides	Macromolecular dimensions	Slow leakage of LDH (M_r 142 000)	Sontag *et al.* (1988)
Streptolysin O	>13 nm	Efflux of urease (M_r 483 000) from sheep erythrocyte membrane vesicles	Buckingham & Duncan (1983)
Patch pipette	Micrometre dimensions	Measurement of tip resistance	Hamill *et al.* (1981), Marty & Neher (1983)

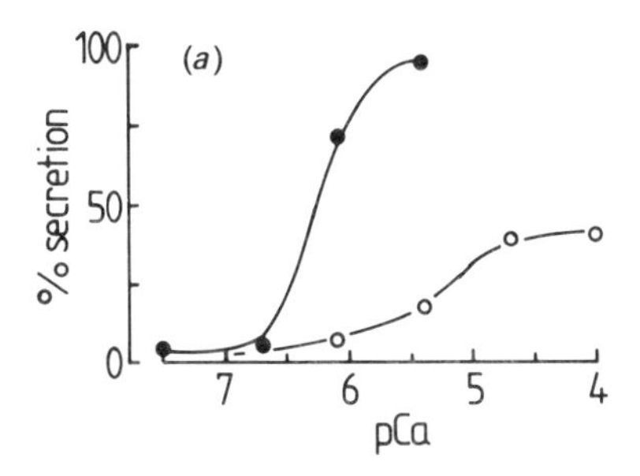

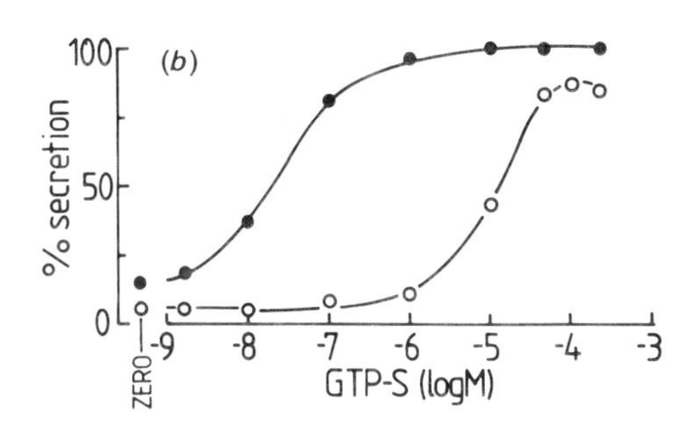

Fig. 2. *Dependence on the concentrations of Ca^{2+} and GTP[S] for secretion from mast cells permeabilized by ATP^{4-} and by streptolysin O*

Mast cells were permeabilized by treatment with either streptolysin O (●) or 30 μM-ATP^{4-} (○) together with (*a*) 15 μM-GTP[S] and a range of pCa or (*b*) pCa 5 and a range of GTP[S]. Incubations were terminated after 30 min. From Koffer & Gomperts (1989).

Another major source of variation concerns the composition of the electrolyte solutions, particularly the anion component, used in permeabilization experiments (Knight & Baker 1982; Sasaki 1984; Churcher & Gomperts, 1990). Here, preference appears to be mainly divided between those who use solutions based on chloride or glutamate. Our own work has mainly concerned the secretory mechanism as it is expressed in rat peritoneal mast cells permeabilized by streptolysin O in chloride-based buffers (Howell & Gomperts, 1987).

Dual Effector System for Exocytosis from Rat Mast Cells

For mast cells a case for a late-acting GTP-binding protein can easily be made. Fig. 3 illustrates the dependence of secretion on three analogues of GTP for cells permeabilized with streptolysin O in the presence of a Ca^{2+} buffer [formulated with EGTA (Tatham & Gomperts, 1990)] to regulate Ca^{2+} at pCa 5 (i.e. 10^{-5} M) (Howell *et al.*, 1987). Before the cells were permeabilized they were treated with metabolic inhibitors to the point of absolute insensitivity to stimulation with exogenous activators, so it can be appreciated that

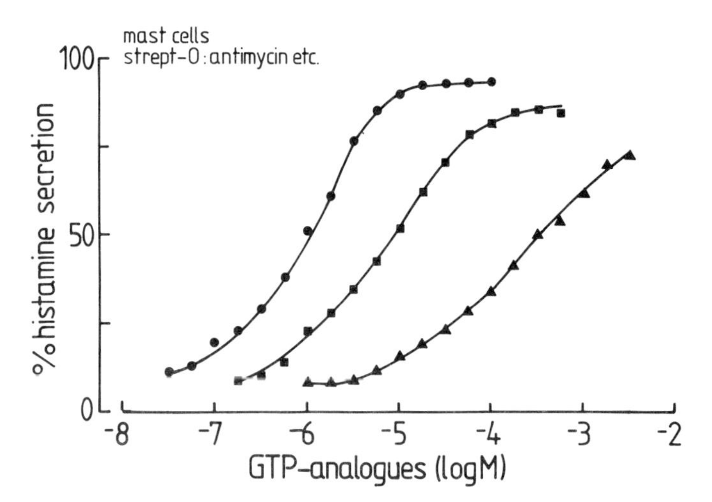

Fig. 3. *Dependence on the concentration of three different analogues of GTP for exocytosis from mast cells permeabilized by streptolysin O*

Mast cells, pretreated with metabolic inhibitors, were permeabilized in the presence of Ca^{2+}/EGTA buffer to regulate pCa 5 and a range of concentrations of GTP[S] (●), Gpp[NH]p (■) and guanosine 5′-[*β*,*γ*-methylene]triphosphate (Gpp[CH_2]p) (▲). Incubations were terminated after 10 min and secreted hexosaminidase present in the supernatants was determined. From Howell *et al.* (1987).

while secretion depends strictly on the presence of a guanine nucleotide, there is no requirement for ATP. Equally (see Fig. 4), secretion depends on the presence of Ca^{2+} at levels elevated above the normal resting concentration of 10^{-7} M (Howell *et al.*, 1987).

It is clear that introduction of guanosine 5′-[*γ*-thio]triphosphate (GTP[S]), or other non-metabolic analogues of GTP, is likely to cause the activation of all GTP-binding proteins present in the cells. However, the consequence of activating G_s/G_i and involvement of adenylyl cyclase can be disregarded both because of the lack of a substrate (ATP) and because any cyclic AMP formed must leak rapidly from streptolysin O-treated cells. Equally, an involvement of G_P-linked phospholipase C is improbable. One of the products of this reaction, inositol trisphosphate (InsP_3), will leak rapidly and any effect that it could have on the level of intracellular Ca^{2+} will be overcome by the (3 mM) Ca^{2+}/EGTA buffer system. Any effect that the other product, diacylglycerol [likely to be retained in the plasma membrane (Bennett *et al.*, 1982)], could have as an activator of protein kinase C is prevented because of the absence of ATP.

Thus, by exclusion, we can conclude that the obligatory role of guanine nucleotides in the exocytotic mechanism of these cells is due to interaction with a GTP-binding protein (G_E) situated close to the exocytotic site. To strengthen the case for the non-involvement of G_P, we have shown that GTP[S]-induced secretion can proceed almost maximally under conditions in which hydrolysis of inositol phospholipids is fully suppressed by neomycin (Cockcroft *et al.*, 1987). As further confirmation of these conclusions, and their relevance to other cell types, it has been shown that in HL-60 cells the rank order of nucleotides for stimulation of phospholipase C and stimulation of exocytosis is different (Stutchfield & Cockcroft, 1988). This shows that in these cells too, the two processes, phospholipase C activation and exocytosis, are also regulated by distinct GTP-binding proteins.

As indicated above, to stimulate exocytosis from permeabilized mast cells all that is necessary is to provide Ca^{2+} and a non-phosphorylating analogue of GTP. Both of these are necessary and together they are sufficient to induce release of the total cellular content of histamine and lysosomal enzymes (which are understood to be contained in the same secretory granules in these cells). This result tells us that there is no obligatory phosphorylation reaction in the terminal stages of the exocytotic pathway. A very similar pattern of control in the exocytotic pathway pertains in guinea-pig eosinophils (Nüsse *et al.*, 1990). The effects of ATP, which will now be described, are those of a modulator, not an effector, of the secretory process.

ATP Modulates Effector Affinity

Fig. 4 illustrates the dependence of secretion on the concentrations of both Ca^{2+} and GTP[S] for cells permeabilized in the absence and presence of ATP (Gomperts & Tatham, 1988). The effect of ATP is simply to enhance the effective affinity for both the essential effectors and we understand that it does this by a phosphorylation reaction mediated by protein kinase C. Thus, inhibition of inositide hydrolysis (by neomycin) with consequent failure to generate diacylglycerol and to activate protein kinase C simply reverses the effect of added ATP (Cockcroft *et al.*, 1987). Inhibitors of protein kinase C also reverse the

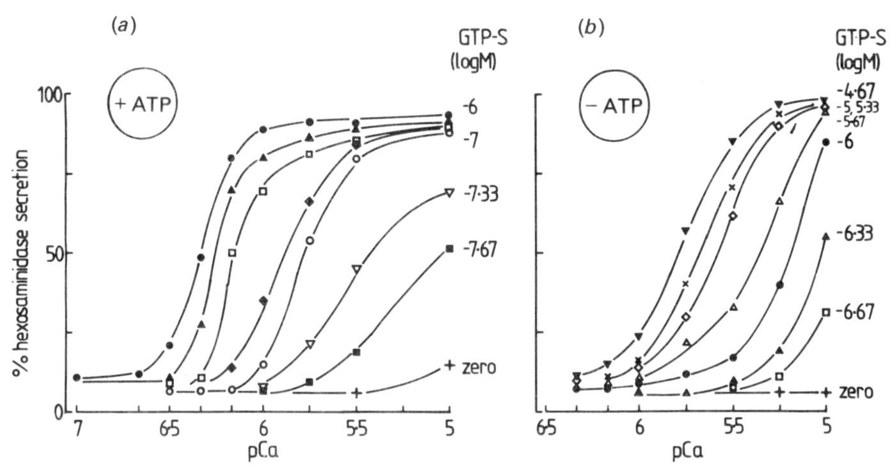

Fig. 4. *Effect of ATP on the concentration dependence of both Ca^{2+} and GTP[S] for exocytosis from mast cells permeabilized by streptolysin O*

Mast cells, pretreated with metabolic inhibitors, were permeabilized (streptolysin O) in the presence (*a*) or absence (*b*) of 1 mM-ATP with GTP[S] and Ca^{2+}/EGTA buffers to regulate pCa as indicated. From Gomperts & Tatham (1988).

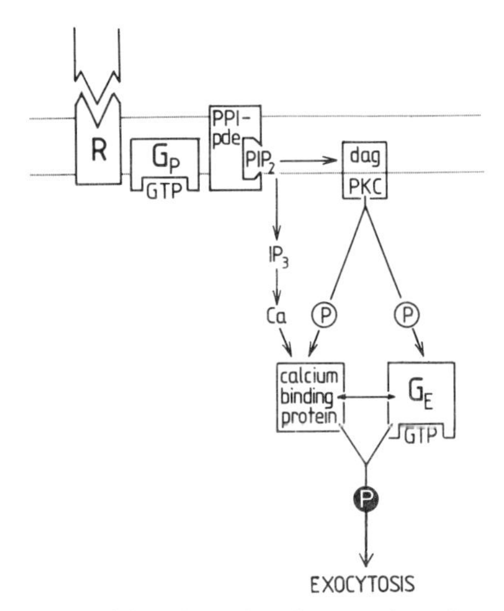

Fig. 5. *Schematic representation of the relationships between the early events in stimulus–secretion coupling and late events mediated by the G-protein, G_E*

Two GTP-binding proteins are understood to be involved in the complete stimulus–secretion pathway. The first of these to be involved, G_P, transduces receptor (R)-mediated signals and activates inositide-specific phosphodiesterase (PPI-pde) to generate InsP_3 (IP_3) and diacylglycerol (dag). In the permeabilized cells, InsP_3 leaks out and control of Ca^{2+} is by the use of Ca^{2+}/EGTA buffers. Diacylglycerol, which is retained in the plasma membrane, can activate protein kinase C if ATP is provided and this enhances the effective affinity for both Ca^{2+} and guanine nucleotide in the exocytotic reaction. Both of these effectors are essential, but elevation of the one reduces the requirement for the other, indicating communication between the Ca^{2+}-binding protein and the G-protein, G_E. ATP exerts an inhibitory action by preserving a control protein in a phosphorylated form so retarding the onset of exocytosis. This can be accelerated by increasing the concentrations of Ca^{2+} and/or GTP[S]. In the absence of this nucleotide, the protein is rapidly dephosphorylated so initiating prompt exocytosis. From Tatham & Gomperts (1989).

effect of added ATP and ensure a low-affinity state for both Ca^{2+} and GTP[S] (Howell *et al.*, 1989). (We have used a diether analogue of diacylglycerol [stearyl methyl glycerol, AMG.C_{16} (Van Blitterswijk *et al.*, 1987; Kramer *et al.*, 1989)] and a pseudosubstrate peptide [PKC-I, RFARKGALRQKNV (House & Kemp, 1987; Alexander *et al.*, 1989)] which can permeate the streptolysin O-treated cells.)

ATP Retards Onset

A case for a dephosphorylation mechanism of exocytosis

In addition to enhancing effector affinity, ATP also modulates the kinetics of exocytosis, but here it is better regarded as an inhibitor (Tatham & Gomperts, 1989; Gomperts & Tatham, 1988). For mast cells permeabilized in the absence of ATP and loaded with GTP[S], exocytosis begins within 3 s of completing the effector pair (i.e. by addition of an activating concentration of Ca^{2+}). By contrast, when ATP is also provided at the time of permeabilization, secretion only begins after a delay, the duration of which is

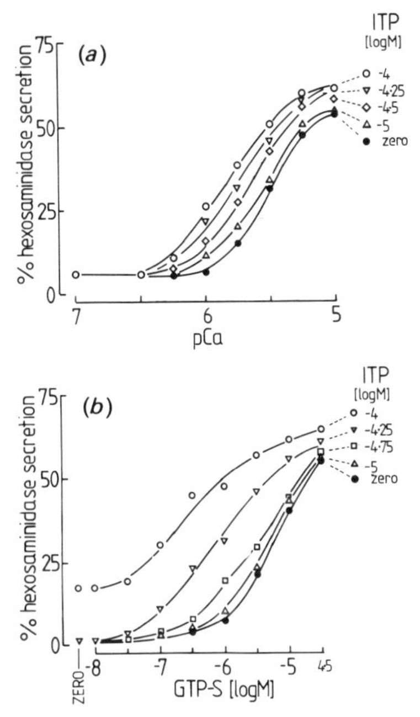

Fig. 6. *Sub-stimulatory concentrations of ITP (10 μM) have the effect of enhancing affinity for GTP[S]*

Mast cells, pretreated with metabolic inhibitors, were permeabilized (streptolysin O) in the presence of (*a*) 3 μM-GTP[S] and Ca^{2+} and ITP as indicated or (*b*) 3 mM-Ca^{2+} buffer to regulate pCa 5.75 with GTP[S] and ITP as indicated.

inversely related to the concentrations of both effectors such that:

$$1/t = k([Ca^{2+}][GTP\ S])^{1/2}$$

We have proposed that there is an enabling reaction leading to the generation of a new steady-state permissive for exocytosis to occur. Since the non-phosphorylating analogue of ATP, adenosine 5′-[$\beta\gamma$-imido]trisphosphate, is without effect (Gomperts & Tatham, 1988), we understand that ATP induces delays by acting as a donor in a phosphorylation reaction. This is unlikely to be mediated by protein kinase C, since pretreatment of the cells with phorbol ester actually accelerates progress through the delay period (Tatham & Gomperts, 1989). The two effectors, Ca^{2+} and guanine nucleotide, which have the effect of reducing the period of delay, thus oppose the effect of phosphorylation. For this reason, we made the suggestion that the effector enzyme of G_E might be a protein phosphatase (Tatham & Gomperts, 1989; Gomperts & Tatham, 1988). This proposal, based on the several interactions of Ca^{2+}, GTP and ATP in the control of exocytosis in mast cells, is presented schematically in Fig. 5.

ITP Activates G_E and Phosphorylates a Subset of Kinase Substrates

A complementary view of the modulation of exocytosis by phosphorylation can be obtained by examining the effects of other nucleotides. All nucleotides capable of stimulating GTP-binding proteins also activate exocytosis (Howell *et al.*, 1987; Churcher *et al.*, 1990). Unlike stimulation by guanine nucleotide analogues, however, ITP-induced secretion is relatively unaffected by the presence or absence of ATP in that the affinity for ITP changes little. This gives a clue that the ITP system is already in a high-affinity state. Support for this idea comes from the finding that secretion induced by ITP can be almost totally suppressed by treating the cells with AMG.C_{16}. This indicates that ITP might itself act as a phosphoryl donor for protein kinase C-mediated phosphorylations involved in the control of effector affinity. In this sense, ITP might act both as a ligand for G_E and also as a substrate for protein phosphorylation. Sub-stimulatory concentrations of ITP (10 μM) have the effect of enhancing affinity for GTP[S] (Fig. 6). However, despite its similarity to ATP in controlling effector affinity, ITP fails to induce delays in the onset of exocytosis, which are only seen if ATP is also included. Here then, we have an activating nucleotide which appears to phosphorylate a subset of proteins which regulate effector affinity, while failing to affect proteins which regulate onset kinetics. It is, however, clearly capable of phosphorylating through the protein kinase C system, corroborating our earlier finding that this is involved in regulation of affinity, but not in the induction of delays.

GTP-Independent Exocytosis

Another case for a dephosphorylation mechanism of exocytosis

Further evidence for a dephosphorylation mechanism and a G-protein-linked protein phosphatase comes from experiments in which the cells are conditioned so as to become susceptible to elevation of cytosol Ca^{2+} alone (Y. Churcher, I. M. Kramer & B. D. Gomperts, unpublished work). The conditioning is carried out by simply withholding the effectors from the cells at the time of permeabilization and then applying them after a few minutes (the permeabilization interval). The argument is based upon the finding that the requirement for the complete dual-effector system (i.e. Ca^{2+} plus GTP[S]) is retained when the cells are similarly conditioned, but in an environment (treatment with 12-*O*-tetradecanoylphorbol 13-acetate or okadaic acid) conducive to maintenance of a high phosphorylation state.

The result illustrated in Fig. 7 shows that the effective affinity for Ca^{2+} (applied together with 10 μM-GTP[S]) in the secretory reaction declines systematically as the permeabilization interval is extended. After about 5 min the affinity has declined to the point at which little secretion occurs even at pCa 5. Restoration of secretory competence and a high affinity for Ca^{2+} can now be secured by addition of ATP (in the range 10–100 μM). A phosphorylation reaction mediated by protein kinase C is again indicated, since recovery does not occur if the cells are treated with specific inhibitors such as AMG.C_{16} (Howell *et al.*, 1989) or PKC-I (Y. Churcher & B. D. Gomperts, unpublished work). It is reasonable to believe that the decline in affinity for Ca^{2+} after permeabilization is a consequence of spontaneous dephosphorylation in the unsupported cells, possibly encouraged by an exodus of protein phosphatase

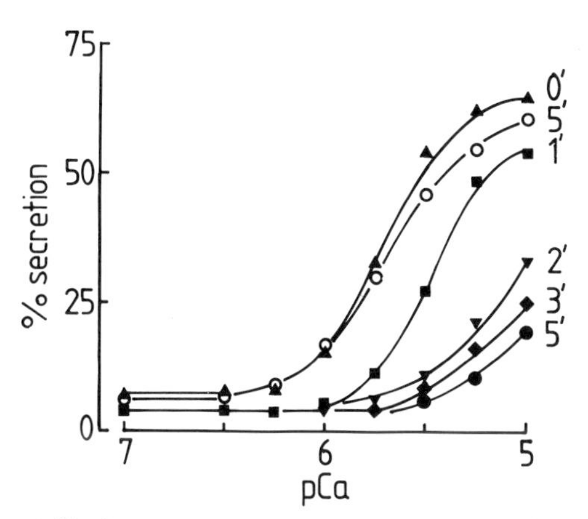

Fig. 7. *Dependence on Ca^{2+} for exocytosis from mast cells stimulated after a permeabilization interval*

Mast cells, pretreated with metabolic inhibitors, were permeabilized (streptolysin O) in the presence of 0.2 mM-Ca^{2+}/EGTA to regulate pCa 7. At the times indicated they were transferred to solutions containing GTP[S] (10 μM) and Ca^{2+} buffers (3 mM) to regulate pCa as indicated in the absence (filled symbols) or presence (open circles, 5 min) of 100 μM-ATP.

inhibitors along with other soluble proteins [e.g. lactate dehydrogenase (Howell & Gomperts, 1987), phosphoglycerate kinase (Gomperts *et al.*, 1987) and actin (Koffer & Gomperts, 1989)] through the induced membrane lesions.

If, instead of considering the dependence on Ca^{2+} for secretion at set times after permeabilization, we measure the dependence on the guanine nucleotide, a very different picture emerges. Fig. 8(*a*) illustrates the dependence on GTP[S] for secretion following its late addition together with Ca^{2+} (pCa 5.25) and the recovery of secretory competence by addition of ATP at 5 min. At this time, instead of simply restoring the affinity for GTP[S], secretion now occurs in its absence. It is as if the guanine nucleotide-dependent step in the exocytotic pathway has been obviated by the simple expedient of conditioning the cells by permeabilization under non-phosphorylating conditions for 5 min. If, as seems reasonable, we apply the same arguments as discussed above concerning spontaneous dephosphorylation during the permeabilization interval, then one possible explanation of this might be that a dephosphorylation reaction, essential for secretion, has occurred during this time so that Ca^{2+} alone is now a sufficient stimulus.

To test this idea we have carried out similar experiments in which the cells were pretreated with okadaic acid [with the aim of suppressing the activities of endogenous protein phosphatases types 1 and 2A (Bialojan & Takai, 1988; Haystead *et al.*, 1989)]. Fig. 8(*b*) illustrates the dependence on GTP[S] for cells pretreated with okadaic acid and then taken through the cycle of metabolic inhibition, pre-permeabilization and restoration of Ca^{2+} affinity with ATP (as in Fig. 8*a*). These cells retain their dependence on the guanine nucleotide.

A similar retention of dependence on guanine nucleotide after restoration of affinity for Ca^{2+} is manifest in cells pretreated with phorbol myristate acetate

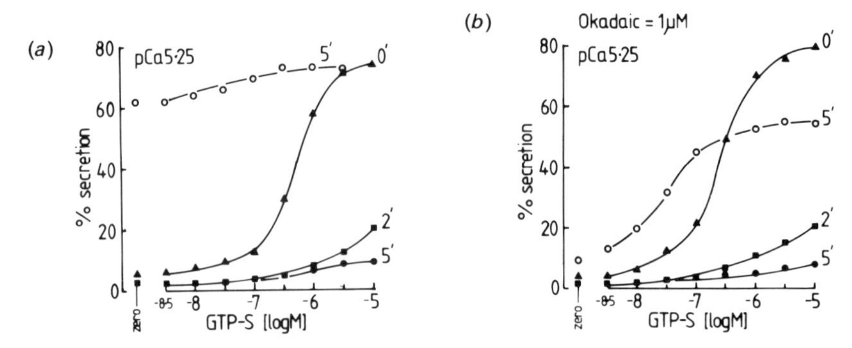

Fig. 8. *Dependence on GTP[S] for exocytosis from mast cells stimulated after a permeabilization interval in the absence (a) and presence (b) of okadaic acid*

Mast cells, pretreated in the presence of dimethylsulphoxide (*a*) or okadaic acid (*b*) and then metabolic inhibitors, were permeabilized (streptolysin O) in the presence of 0.2 mM-Ca^{2+}/EGTA to regulate pCa 7. At the times indicated, they were transferred to solutions containing 3 mM-Ca^{2+}/EGTA buffer to regulate pCa 5.25, and a range of GTP[S] as indicated, in the absence (filled symbols) or presence (open circles) of 100 μM-ATP.

(not shown). These results strongly support the conclusion that protein dephosphorylation comprises a control step in the terminal pathway of regulated exocytosis. A corollary of this is that the effector of the GTP-binding protein (G_E) mediating regulated exocytosis is likely to be a protein phosphatase.

Two Dephosphorylation Reactions in the Terminal Pathway of Exocytosis?

We should consider whether the action of guanine nucleotide which accelerates the silent reaction preceding exocytosis from cells permeabilized in the presence of ATP is identical with that revealed by treating cells with okadaic acid or phorbol ester in the manner just described? If the two reactions (presumed to be dephosphorylations) were one and the same, then we might expect that the protein phosphatase inhibitor okadaic acid would induce delays similar to ATP. Despite many attempts, we have never been able to detect any effect of okadaic acid pretreatment on the kinetics of exocytosis. In addition, the effect of pretreating cells with phorbol ester is actually to accelerate, not to delay, the onset of exocytosis (Tatham & Gomperts, 1989). While the interpretation of this latter observation is made more complicated owing to the enhancement by protein kinase C activation of the effective affinity for the two stimulating effectors, these results render it unlikely that the products of protein kinase C-mediated phosphorylations form the impediment to immediate onset of exocytosis. We thus conclude that the two steps tentatively identified as protein dephosphorylation reactions, the one perceived kinetically, the other by the use of compounds such as okadaic acid, are quite distinct. (An alternative view which cannot be excluded until biochemical data are to hand, is that our original argument, concerning a phosphorylation mechanism for the induction of delays by ATP, was mistaken.)

Is it reasonable to believe that there are two sites of action of GTP in the last steps of the exocytotic process? There are of course no direct data.

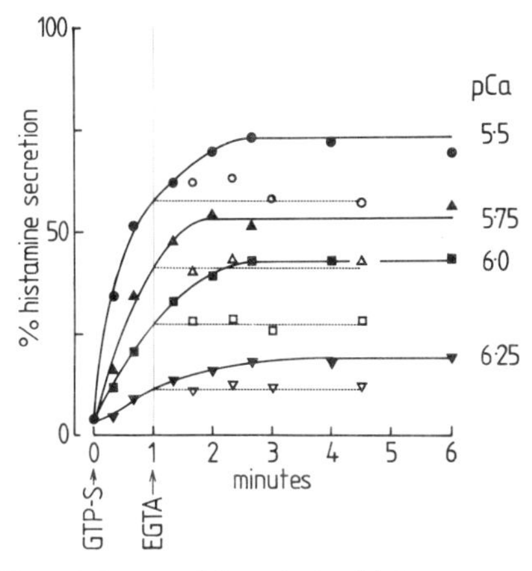

Fig. 9. *Effect of abrupt withdrawal of Ca^{2+} from exocytosing cells*

Mast cells were permeabilized in the presence of GTP[S] (10 μM) in the presence of 0.2 mM-Ca^{2+}/EGTA to regulate pCa 7 and 1 mM-ATP. After 2 min, the cells were transferred to solutions containing 3 mM-Ca^{2+}/EGTA buffers (regulating pCa 6.25, 6, 5.75 and 5.5 as indicated) to initiate secretion and timed samples were removed and quenched. After 1 min, portions of the secreting cells were transferred to 5 mM-EGTA and sampling was continued. The intercepts between the progress curves and the vertical dotted line indicates the extent of secretion which had occurred at the time of addition of EGTA. From Gomperts *et al.* (1988).

However, there is some evidence that the other essential effector, Ca^{2+}, does interact twice in the terminal stages. The first site of action of Ca^{2+} determines the rate of progression through the silent reaction as already described. Yet even when exocytosis has started, the requirement for Ca^{2+} remains, since if it is rapidly withdrawn (by addition of excess EGTA) then secretion ceases abruptly as shown in Fig. 9. Unfortunately, the question of whether GTP strikes twice is not amenable to testing in this manner, since guanine nucleotides cannot be abruptly removed.

Protein Dephosphorylation in the Mechanism of Exocytosis

It may seem surprising that exocytosis can be activated in conditions in which the cells are totally depleted of phosphorylating nucleotides and there is no possibility of phosphorylation. In contrast, application of metabolic inhibitors to intact cells renders them totally unresponsive to all forms of receptor-mediated stimulation and also to Ca^{2+} ionophores. The requirement for metabolic competence [generally taken as a requirement for intracellular ATP (Johansen, 1987)] can readily be understood on the basis of (i) the need to maintain the level of phosphatidylinositol 4,5-bisphosphate which is the source of InsP_3 [mobilizes intracellular Ca^{2+} (Berridge, 1987)] and diacylglycerol [activator of protein kinase C (Berridge, 1987) which regulates the affinity of the effector systems]; (ii) the need to maintain the level of GTP which is the ligand for activation of G_P-linked phospholipase C (Cockcroft & Gomperts, 1985) and (iii) GTP again, which is required for the activation of G_E [this probably explains why Ca^{2+} ionophore-induced secretion is suppressed by metabolic depletion (Foreman *et al.*, 1973; Bennett *et al.*, 1980)]. None of these constraints applies in permeabilized cells, in which the initial events related to inositide metabolism are obviated, and in which Ca^{2+} and GTP (analogues) can be provided.

While phosphorylation does not comprise an essential step in the terminal stages of the exocytotic pathway, is it really realistic to believe that the reverse reaction, protein dephosphorylation, might be involved? There are some indications from other systems that this might be the case. The best evidence for protein dephosphorylation in the terminal pathway of exocytosis is provided by the ciliated protozoan *Paramecium tetraurelia* (Plattner, 1987). In this large unicellular organism, a synchronous exocytotic process, the trichocyst discharge reaction, occurs within 1 s of appropriate stimulation (e.g. by application of aminoethyl dextran). Simultaneous with this, one phosphoprotein present in resting cells, undergoes dephosphorylation (Zieseniss & Plattner, 1985). Non-discharge mutants (nd9) which are unable to dephosphorylate this protein fail to respond (Zieseniss & Plattner, 1985) indicating that the apo-form must be generated in order to allow progress along the exocytotic pathway (Stecher *et al.*, 1987). Exocytosis can be initiated by direct micro-injection of alkaline phosphatase (Momayezi *et al.*, 1987) and the receptor-stimulated process can be inhibited by injection of antibodies to calcineurin (Momayezi *et al.*, 1987) (itself a protein phosphatase; Stewart *et al.*, 1982). All this adds up to very strong direct evidence for a dephosphorylation pathway of exocytosis in this organism. The scope of our experiments on mammalian cells lacks the genetic dimension which gives such strength to the conclusions relating to the lower organism and our approach has necessarily been at the physiological rather than the biochemical level of definition. However, proteins having immunological cross-reactivity with the exocytosis-related *Paramecium* phosphoprotein appear to be widespread in the secretory tissues of mammals (Satir *et al.*, 1989 Stecher *et al.* 1987).

What Kind of GTP-Binding Protein is G_E ?

Here we have been considering the role(s) of GTP-binding proteins in the stimulation of regulated secretion. GTP-binding proteins have also been implicated as mediators of vesicle fusion to enable trafficking to occur between the endoplasmic reticulum and the Golgi stacks (Beckers & Balch, 1989), and then through the Golgi (Segev *et al.*, 1988; Melançon *et al.*, 1987) and again in the terminal stages of the constitutive secretory pathway in cells as diverse as yeast (Salminen & Novick, 1987; Novick *et al.*, 1988) and mammalian liver (Melançon *et al.*, 1987). We suspect that the structural and mechanistic interactions of GTP-binding proteins in the constitutive and regulated pathways of exocytosis are quite distinct. In contrast to the regulated exocytotic process of mast cells discussed here, in constitutive exocytosis the requirement is for GTP. Its non-hydrolysable analogues actually inhibit membrane fusion (Melançon *et al.*, 1987). From this it is understood that the GTP-binding proteins involved in vesicle trafficking and constitutive exocytosis must cycle repeatedly between the GTP- and GDP-bound states (Novick *et al.*, 1988; Bourne, 1988). In non-secreting mutants of *Saccharomyces cerevisiae*, altered forms of two monomeric (i.e. roughly ras-like) GTP-binding proteins associated with blockade of the pathway at an early stage (YPT-1; Segev *et al.*, 1988) and in the terminal stages (sec4; Salminen & Novick 1987; Novick *et al.*, 1988) have been identified. For the stimulation of regulated exocytosis, the effective affinity for the non-hydrolysable analogues is as much as 10^3 times higher than that of the parent nucleotide (Howell *et al.*, 1987) and so it is likely that the better stimulus occurs as a result of persistent activation. A more suitable paradigm for the structure and mechanism of the exocytosis-related G_E proteins might therefore be provided by the cell-surface receptor linked heterotrimeric forms of GTP-binding proteins in which activation leads to dissociation and release of the α-subunit.

Work in our laboratory has been financed by project grants from the Wellcome Trust. Further generous support has been provided by The Nuffield Foundation, The Vandervell Trust and the Gower Street Secretory Mechanisms Group. We are most grateful to Dr Y. Tsukitani (Fujisawa Pharmaceutical Co.) for the gift of okadaic acid and to Dr Dennis Alexander for PKC-I peptide.

References

Alexander, D. R., Hexham, J. M., Lucas, S. C., Graves, J. D., Cantrell, D. A. & Crumpton, M. J. (1989) *Biochem. J.* **260**, 893–901
Barrowman, M. M., Cockcroft, S. & Gomperts, B. D. (1987) *J. Physiol. (London)* **383**, 115–124
Beaven, M. A. & Cunha-Melo, J. R. (1988) *Prog. Allergy* **42**, 123–184
Beaven, M. A., Guthrie, D. F., Moore, J. P., Smith, G. A., Hesketh, T. R. & Metcalfe, J. C. (1987) *J. Cell. Biol.* **105**, 1129–1136
Beckers, C. J. M. & Balch, W. E. (1989) *J. Cell. Biol.* **108**, 1245–1256
Bennett, J. P., Cockcroft, S. & Gomperts, B. D. (1980) *Nature (London)* **282**, 851–853
Bennett, J. P., Cockcroft, S. & Gomperts, B. D. (1981) *J. Physiol. (London)* **317**, 335–345
Bennett, J. P., Cockcroft, S., Caswell, A. H. & Gomperts, B. D. (1982) *Biochem. J.* **208**, 801–808
Berridge, M. J. (1987) *Annu. Rev. Biochem.* **56**, 159–193
Bialojan, C. & Takai, A. (1988) *Biochem. J.* **256**, 283–290
Bittner, M. A., Holz, R. W. & Neubig, R. R. (1986) *J. Biol. Chem.* **261**, 10182–10188
Bourne, H. R. (1988) *Cell (Cambridge, Mass.)* **53**, 669–671
Brown, E. M., LeBoff, M. S., Oetting, M., Possillico, J. T. & Chen, C. (1987) *Recent Prog. Horm. Res.* **43**, 337–382
Buckingham, L. & Duncan, J. L. (1983) *Biochim. Biophys. Acta.* **729**, 115–122
Cheek, T. R. (1989) *J. Cell Sci.* **93**, 211–216
Choy, V. J. & Watkins, W. B. (1976) *J. Endocrinol.* **69**, 349–358
Churcher, Y. & Gomperts, B. D. (1990) *Cell Regul.* **1**, 337–346
Churcher, Y., Allan, D. & Gomperts, B. D. (1990) *Biochem. J.* **266**, 157–163
Cockcroft, S. & Gomperts, B. D. (1979) *Nature (London)* **279**, 541–542
Cockcroft, S. & Gomperts, B. D. (1985) *Nature (London)* **314**, 534–536
Cockcroft, S., Howell, T. W. & Gomperts, B. D. (1987) *J. Cell. Biol.* **105**, 2745–2750
Dormer, R. L. & Ashcroft, S. J. H. (1974) *Biochem. J.* **144**, 543–550
Douglas, W. W. (1974) *Biochem. Soc. Symp.* **39**, 1–28
Foreman, J. C., Mongar, J. L. & Gomperts, B. D. (1973) *Nature (London)* **245**, 249–251
Fray, J. C. S., Park, C. S. & Valentine, A. N. D. (1987) *Endocr. Rev.* **8**, 53–93
Füssle, R., Bhakdi, S., Sziegoleit, A., Tranum-Jensen, J., Kranz, T. & Wellensiek, H. J. (1981) *J. Biol. Chem.* **91**, 83–94
Gomperts, B. D. (1989) in *G-Proteins* (Iyengar, R. & Birnbaumer, L., eds.), pp. 601–637, Academic Press, San Diego
Gomperts, B. D. (1990) *Annu. Rev. Physiol.* **52**, 591–606
Gomperts, B. D. & Fernandez, J. M. (1985) *Trends Biochem. Sci.* **10**, 414–417

Gomperts, B. D. & Tatham, P. E. R. (1988) *Cold Spring Harbor Symp. Quant. Biol.* **53**, 983–992
Gomperts, B. D., Cockcroft, S., Howell, T. W., Nüsse, O. & Tatham, P. E. R. (1987) *Biosci. Rep.* **7**, 369–381
Gomperts, B. D., Cockcroft, S., Howell, T. W. & Tatham, P. E. R. (1988) in *Molecular Mechanisms in Secretion* (Thorn, N. A., Treiman, M. & Petersen, O. H., eds.), pp. 248–261, Munksgaard, Copenhagen
Hamill, O. P., Marty, A., Neher, E., Sakman, B. & Sigworth, F. J. (1981) *Eur. J. Physiol.* **391**, 85–100
Handwerger, S., Conn, P. M., Barrett, J., Barry, S. & Golander, A. (1981) *Am. J. Physiol.* **240**, E550–E555
Haslam, R. J. & Davidson, M. M. L. (1984) *FEBS Lett.* **174**, 90–95
Haystead, T. A. J., Sim, A. T. R., Carling, D., Honnor, R. C., Tsukitani, Y., Cohen, P. & Hardie, D. G. (1989) *Nature* (*London*) **337**, 78–81
House, C. & Kemp, B. E. (1987) *Science* **238**, 1726–1728
Howell, T. W. & Gomperts, B. D. (1987) *Biochim. Biophys. Acta.* **927**, 177–183
Howell, T. W., Cockcroft, S. & Gomperts, B. D. (1987) *J. Cell. Biol.* **105**, 191–197
Howell, T. W., Kramer, Ij. & Gomperts, B. D. (1989) *Cellular Signalling* **1**, 157–163
Impraim, C. C., Foster, K. A., Micklem, K. J. & Pasternak, C. A. (1980) *Biochem. J.* **186**, 847–860
Johansen, T. (1987) *Pharmacol. Toxicol.* **61** (Suppl II), 1–20
Johnson, G. S. & Mukku, V. R. (1979) *J. Biol. Chem.* **254**, 95–100
Knight, D. E. & Baker, P. F. (1982) *J. Membr. Biol.* **68**, 107–140
Knight, D. E. & Baker, P. F. (1985) *FEBS Lett.* **189**, 345–349
Knight, D. E. & Scrutton, M. C. (1986) *Eur. J. Biochem.* **160**, 183–190
Koffer, A. & Gomperts, B. D. (1989) *J. Cell. Sci.* **94**, 585–591
Kramer, Ij. M., Van Der Bend, R. L., Tool, A. T. J., Van Blitterswijk, W. J., Roos, D. & Verhoeven, A. J. (1989) *J. Biol. Chem.* **264**, 5876–5884
Leslie, B. A., Putney, J. W. & Sherman, J. M. (1976) *J. Physiol.* (*London*) **260**, 351–370
Marquardt, D. L., Gruber, H. E. & Walker, L. L. (1987) *J. Pharmacol. Exp. Ther.* **240**, 145–149
Marty, A. & Neher, E. (1983) in *Single Channel Recording* (Sakmann, B. & Neher, E., eds.), pp. 107–121, Plenum, New York
McMillian, M. K., Soltoff, S. P. & Talamo, B. R. (1988) *Biochem. Pharmacol.* **37**, 3790–3793
Melançon, P., Glick, B. S., Malhotra, V., Weidman, P. J., Serafini, T., Gleason, M. L., Orci, L. & Rothman, J. E. (1987) *Cell* (*Cambridge, Mass.*) **51**, 1053–1062
Miller, M. R., Castellot, J. J. & Pardee, A. B. (1978) *Biochemistry* **17**, 1073–1080
Momayezi, M., Lumpert, C. J., Kersken, H., Gras, U., Plattner, H., Krinks, M. H. & Klee, C. B. (1987) *J. Cell. Biol.* **105**, 181–189
Nemeth, E. F. & Scarpa, A. (1986) *FEBS Lett.* **203**, 15–19
Novick, P. J., Goud, B., Salminen, A., Walworth, N. C., Nair, J. & Potenza, M. (1988) *Cold Spring Harbor Symp. Quant. Biol.* **53**, 637–647
Nüsse, O., Lindau, M., Cromwell, O., Kay, A. B. & Gomperts, B. D. (1990) *J. Exp. Med.* **171**, 775–786
Oetting, M., Leboff, M. S., Levy, S., Swiston, L., Preston, J., Chen, C. & Brown, E. M. (1987) *Endocrinology* **121**, 1571–1576
Park, C. S., Honeyman, T. W., Chung, E. S., Lee, J. S., Sigmon, D. H. & Fray, J. C. S. (1986) *Am. J. Physiol.* **251**, F1055–F1062
Penner, R., Pusch, M. & Neher, E. (1987) *Biosci. Rep.* **7**, 313–321
Plattner, H. (1987) in *Cell Fusion* (Sowers, A. E., ed.), pp. 69–98, Plenum, New York
Salminen, A. & Novick, P. J. (1987) *Cell* (*Cambridge, Mass.*) **49**, 527–538
Sasaki, H. (1984) *Dev. Biol.* **101**, 125–135
Satir, B. H., Hamasaki, T., Reichman, M. & Murtaugh, T. J. (1989) *Proc. Natl. Acad. Sci. U.S.A.* **86**, 930–932
Segev, N., Mulholland, J. & Botstein, D. (1988) *Cell* (*Cambridge, Mass.*) **52**, 915–924
Sherwood, L. M., Potts, J. T., Care, A. D., Mayer, G. P. & Aurbach, G. D. (1966) *Nature* (*London*) **209**, 52–55
Shoback, D. M., Thatcher, J. G., Leombruno, R. & Brown, E. M. (1983) *Endocrinology* **113**, 424–426
Sontag, J.-M., Aunis, D. & Bader, M.-F. (1988) *Eur. J. Cell Biol.* **46**, 316–326
Stecher, B., Höhne, B., Gras, U., Momayezi, M., Glas-Albrecht, R. & Plattner, H. (1987) *FEBS Lett.* **223**, 25–32
Stewart, A. A., Ingebritsen, T. S., Manalan, A., Klee, C. B. & Cohen, P. (1982) *FEBS Lett.* **137**, 80–84
Stutchfield, J. & Cockcroft, S. (1988) *Biochem. J.* **250**, 375–382

Tatham, P. E. R. & Gomperts, B. D. (1989) *Biosci. Rep.* **9**, 99–109
Tatham, P. E. R. & Gomperts, B. D. (1990) in *Peptide Hormones — A Practical Approach* (Siddle, K. & Hutton, J. C., eds.) IRL Press, Oxford, in the press
Tatham, P. E. R. & Lindau, M. (1990) *J. Gen. Physiol.* **95**, 459–476
Tsien, R. Y. (1986) in *Optical Methods in Cell Physiology* (de Weer, P. & Salzberg, B. M., eds.), pp. 327–346, John Wiley, New York
Van Blitterswijk, W. J., Van Der Bend, R. L., Kramer, Ij., Verhoeven, A. J., Hilkmann, H. & De Widt, J. (1987) *Lipids* **22**, 842–846
Wilson, B., Deanin, G., Stump, R. & Oliver, J. (1988) *FASEB J.* **2**, A1236
Zieseniss, E. & Plattner, H. (1985) *J. Cell Biol.* **101**, 2028–2035

Biochem. Soc. Symp. **56**, 103–116
Printed in Great Britain

The Role of *ras* Gene Products in Second Messenger Generation

MICHAEL J. O. WAKELAM, FIONA M. BLACK and SHIREEN A. DAVIES

Molecular Pharmacology Group, Department of Biochemistry, University of Glasgow, Glasgow G12 8QQ, Scotland, U.K.

Introduction

The human *ras* family comprises three similar genes, Ha-, N- and Ki-. They were first characterized as viral oncogenes carried by two related transforming retroviruses. These were the Harvey Murine Sarcoma virus (v-Ha-*ras*) and the Kirsten Murine Sarcoma virus (v-Ki-*ras*) (see Barbacid, 1987). The N-*ras* gene was identified as the transforming gene in a human neuroblastoma (Barbacid, 1987). Some 30% of human tumours have been estimated to have an activated *ras* oncogene, while approximately 1% demonstrate an over-expression of a normal proto-oncogenic *ras*.

The *ras* genes each encode a 21 kDa protein, $p21^{ras}$, of 189 amino acids, except in the case of Ki-*ras* where there are only 188. A great deal of sequence similarity exists between the three gene products. Within amino acids 1–164 there are a maximum of 15 differences, 8 of which are found between amino acids 121 and 135. However, between amino acids 165 and 185 there is almost no sequence similarity. Beyond 185 is a conserved CAAX box, the cysteine residue of which is acylated. The acylation of the $p21^{ras}$ proteins is essential both for their plasma membrane association and their biological activity. It was previously thought that Cys-186 was palmitoylated, but recent evidence from Hancock *et al.* (1989) has demonstrated that this amino acid is in fact polyisoprenylated and carboxyl-methylated, the three downstream amino acids being removed by proteolysis. The Cys-181 and Cys-184 residues in $p21^{Ha\text{-}ras}$ and Cys-181 in $p21^{N\text{-}ras}$ are then also palmitoylated; $p21^{Ki\text{-}ras}$, which does not have these upstream Cys residues, is not palmitoylated.

$p21^{ras}$ proteins bind guanine nucleotides. When isolated from *Escherichia coli* expression systems the proteins have one mole of GDP bound per mole of $p21^{ras}$ (Hall & Self, 1986). The GDP and GTP dissociation constants are $\sim 2 \times 10^{-11}$ M (Feursten *et al.*, 1987), with the rate-determining step in nucleotide exchange being the nucleotide off-rate, thus exchange is pseudo first-order (Hall & Self, 1986). In addition to binding guanine nucleotides, the $p21^{ras}$ proteins exhibit an intrinsic GTPase activity (see below).

In human tumours, activating mutations have been detected at codons 12, 13 and 61 (Bos, 1988). Analysis of the possible mutations at codon 12 has demonstrated that the glycine residue can be replaced by any other amino acid with the exception of proline to generate a transforming protein (Seeburg *et al.*, 1984), while at codon 61 replacement of the glutamine by any amino acid other than glutamic acid or proline also generates oncogenic *ras* (Der *et al.*, 1986). Since these mutations are in the regions of $p21^{ras}$ thought to be involved in guanine nucleotide binding (see DeVos *et al.*, 1988), it was predicted that these mutations would induce transformation as a consequence of a major reduction in intrinsic GTPase activity. This prediction was questioned since *in vitro* the activating mutations only induced small reductions in GTPase activity. The explanation of this apparent paradox came with the discovery of GTPase-activating protein (GAP) (Trahey & McCormick, 1987). GAP was found to increase the GTPase activity of normal N-*ras* more than 200-fold *in vitro*, but had no effect upon Asp-12 or Val-12 mutants (Trahey & McCormick, 1987). The cDNA for bovine GAP has been cloned generating a protein of molecular mass of 125 000 kDa with 1044 amino acids (Vogel *et al.*, 1988). Two models for the function of GAP in the mechanism of action of $p21^{ras}$ have been proposed. In the first model, GAP is a $p21^{ras}$-activated effector protein; in the second, GAP regulates the function of the p21 protein by stimulating its GTPase activity and promoting the generation of the 'inactive' GDP-bound form. Recent evidence from Ballester *et al.* (1989) that mammalian GAP can function in yeast cells to regulate the GTPase activity of yeast RAS and act as a feedback regulator of RAS function would not support the idea that GAP is an effector molecule.

The $p21^{ras}$ proteins are thus membrane bound, bind guanine nucleotides and express an intrinsic GTPase activity. Consequently, they have been proposed to function as guanine-nucleotide-binding regulatory proteins (G-proteins) in cells. These may function to couple the receptors for certain growth factors to the generation of intracellular messages which may be involved in the regulation of cell proliferation (Barbacid, 1987). The signal transduction pathways implicated in the regulation of cell proliferation are those involving adenylate cyclase, phosphoinositidase C, phosphospholipase D and tyrosine kinase activity (see Milligan & Wakelam, 1989 for a review). The work described in this chapter focuses upon investigations into the effect of $p21^{ras}$ upon these signalling pathways.

Cyclic AMP Metabolism

The role of cyclic AMP in the regulation of cell proliferation is controversial. Both stimulatory and inhibitory effects have been described. In yeast cells RAS proteins control the activity of adenylate cyclase (Toda *et al.*, 1985). We have, therefore, examined the regulation of cyclic AMP metabolism in NIH-3T3 fibroblasts transformed by the over-expression of $p21^{N\text{-}ras}$ (T15 cells). The T15 cell line contains a human fetal genomic N-*ras* gene under the control of the steroid-inducible mouse mammary tumour virus–long terminal repeat (MMTV–LTR) promotor (Wakelam *et al.*, 1986); addition of dexamethasone to the culture medium induces the expression of the *ras* gene and results in the onset of transformation of the cells. T15 cells were cultured under three separate conditions before cyclic AMP metabolism was examined: (i) in the absence of the inducer dexamethasone (T15 −); (ii) in the presence of 2 μM-dexamethasone for 24 h before experiments (T15 ±) and (iii) in the presence of the inducer for at least two passages (T15 +) before the experiments. Adenylate cyclase activity was stimulated in the cells either with the β-adrenergic receptor agonist, isoprenaline, or with the diterpene forskolin, which directly activates the enzyme. Experiments using the β-adrenergic receptor antagonist, propranolol, demonstrated the absence of α_2-adrenergic receptors on NIH-3T3 cells or their *ras*-transformed derivatives (Davies *et al.*, 1989).

Expression of the *ras* gene in fibroblasts caused a small but non-significant reduction in resting intracellular cyclic AMP concentration. However, there was a reduction in the ability of β-adrenergic stimulation to increase cyclic AMP levels within the cells. Fig. 1 shows that this reduction is apparent in the T15 ± cells, but that in the fully transformed T15 + cells agonist-stimulated cyclic AMP generation was only about 25% of that observed in the control, untransformed T15 − cells. Two separate mechanisms appear to be responsible for this reduction. First, there appears to be a loss in adenylate cyclase activity in the fully transformed T15 + cells, since forskolin-stimulated cyclic AMP generation is only 25% of that in both the T15 − and the T15 ± cells (Fig. 1). Secondly, expression of the *ras* gene in the cells induces a loss in the number of β-adrenergic receptors on the cell surface (Fig. 2). Examination of another cell line (N866), which constitutively over-expresses p21$^{N\text{-}ras}$, demonstrated that this reduction in both forskolin- and agonist-stimulated cyclic AMP generation appears to be a general effect of over-expression of this *ras* gene in fibroblasts (Davies *et al.*, 1989), indeed in this cell line the loss of β-adrenergic-stimulated adenylate cyclase activity was total. The reduced basal concentra-

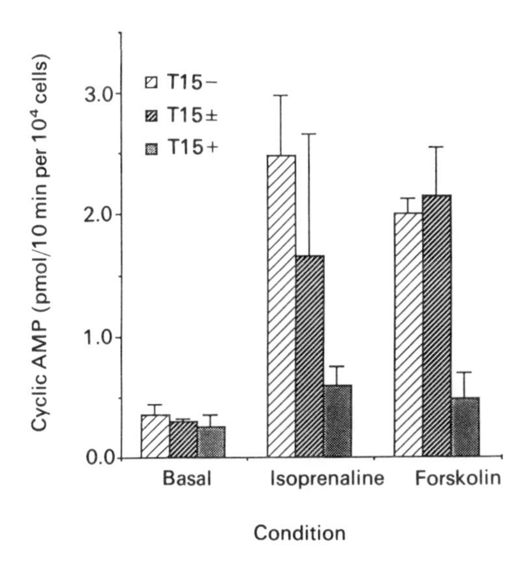

Fig. 1. *Stimulation of cyclic AMP generation in ras-transformed cells*

Cyclic AMP generation was determined as described by Davies *et al.* (1989) in control (T15 −), 24 h *ras* (T15 ±) and fully *ras*-transformed (T15 +) cells. Isoprenaline concentration was 10^{-5} M and that of forskolin 10^{-4} M.

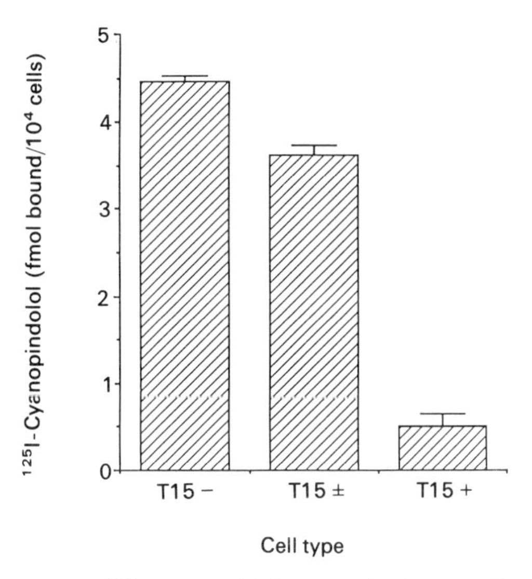

Fig. 2. *Binding of ^{125}I-cyanopindolol to control and ras-transformed cells*
Maximum binding experiments were performed on whole cells as described in Davies *et al.* (1989).

tions of cyclic AMP in the transformed cells could be due to an increase in an inhibitory G-protein input, indeed Franks *et al.* (1987) have suggested that $p21^{Ki\text{-}ras}$ affects adenylate cyclase activity by interacting with G_i. We have found that expression of $p21^{N\text{-}ras}$ in the T15 cell has no effect upon the levels of G_s, G_i or G_o (Milligan *et al.*, 1989). Additionally, if T15 cells are treated with 100 ng of pertussis toxin/ml for 16 h to completely ADP-ribosylate the available G_i in the cells, no effect is observed upon basal or agonist-stimulated cyclic AMP generation (Davies *et al.*, 1989); thus, expression of the *ras* gene does not cause an increase in the input of G_i to adenylate cyclase activity.

In *ras*-transformed cells there is an increase in the activity of protein kinase C (see below). This enzyme can phosphorylate β-receptors and receptor phosphorylation precedes sequestration, loss from the cell surface and degradation of the receptors (Sibley *et al.*, 1987) and thus this may explain the reduction in β-receptor number (Fig. 2). The reduction in adenylate cyclase activity appears to be a consequence of transformation since a similar effect was observed in c-*sis*-transformed cells (Davies *et al.*, 1989).

The importance of the reduced responsiveness of the *ras*-transformed cells to agents which raise intracellular cyclic AMP is emphasized by experiments where the intracellular concentration of the nucleotide is increased by the inclusion of increasing concentrations of dibutyryl cyclic AMP in the culture medium. Millimolar medium concentrations of the agent cause approximately 50% inhibition of proliferation of NIH-3T3 and T15− cells, on the other hand, the same concentration of dibutyryl cyclic AMP causes the complete inhibition of proliferation of T15+ cells within 24 h (Davies *et al.*, 1989). Indeed, Tagliaferri *et al.* (1988) have demonstrated that site-selective cyclic

AMP analogues can reverse the transformation of NIH-3T3 cells induced by v-Ha-*ras*.

Inositol Phospholipid Metabolism

A considerable number of reports have suggested that activation of the inositol phospholipid pathway is fundamental to the mechanism whereby certain growth factors, in particular peptide mitogens, stimulate cell proliferation (see Berridge, 1987, for a review). The basis of this proposal is that activation of phosphatidylinositol 4,5-bisphosphate [PtdIns(4,5)P_2] hydrolysis generates inositol 1,4,5-trisphosphate [Ins(1,4,5)P_3] and *sn*-1,2-diacylglycerol (DAG) as products. These two molecules are both second messengers, with Ins(1,4,5)P_3 stimulating the release of intracellular stored Ca^{2+} and DAG activating protein kinase C (see Berridge & Irvine, 1989, for a review). Stimulation of this enzyme and an increase in intracellular free Ca^{2+} concentration are two fundamental events in mitogen-stimulated proliferation (see Milligan & Wakelam, 1989). Agonist-stimulated PtdIns(4,5)P_2 hydrolysis involves an as yet uncharacterized G-protein (G_p), which exists in at least two forms, one of which is sensitive to pertussis toxin treatment of cells (Boyer *et al.*, 1989). Since $p21^{ras}$ appears to be a G-protein, and stimulated inositol lipid turnover has been suggested to be critical for mitogenesis, considerable effort has been extended into an investigation into whether $p21^{ras}$ either is, or can function as, G_p. In this model, the transforming mutations of $p21^{ras}$, which have a reduced intrinsic GTPase activity, would be constitutively active and thus, if coupled to phosphoinositidase C, bring about constitutive activation of the enzyme and lead to an increase in PtdIns(4,5)P_2 hydrolysis in the absence of receptor stimulation.

Fleischman *et al.* (1986) detected an increase in the turnover of inositol phospholipids in both NRK and NIH-3T3 cells transformed by three different activated *ras* genes. Kamata *et al.* (1987) demonstrated that NIH-3T3 cells transformed by the v-Ki-*ras* gene also display elevated inositol phospholipid turnover, while in a flat, untransformed *ras*-resistant variant, derived from the same transfection, no differences in inositol phospholipid turnover from the untransformed parental NIH-3T3 cells were detected.

We have investigated a role for $p21^{ras}$ in inositol lipid metabolism by using two experimental approaches. In the first, NIH-3T3 cells which have been transfected with normal or mutated *ras* genes have been used. These cells are transformed as a consequence of at least a 10-fold over-expression of the normal genes, either constitutively, or in two cell lines, T15 and H8/22, by dexamethasone induction of expression of N- or Ha-*ras*, respectively, due to the genes being under the control of the steroid-inducible MMTV–LTR promotor. In the cells transformed by mutant *ras* genes, the gene product is not over-expressed, but is rather mutated at codons 12, 13, or 61. Experiments were also performed with the pLHT5.11 cell, where expression of the Lys-61 mutation in N-*ras* was again regulated by the MMTV–LTR promotor. The second approach involved transient transfections of COS-1 cells with normal and mutated Ha-*ras* genes.

Over-expression of the normal N-*ras* proto-oncogene in the T15 cell line resulted in an increase in the stimulation of inositol phosphate generation in response to a number of mitogens (Wakelam *et al.*, 1986). This was particularly apparent for bombesin. The amplification appeared to be dependent upon the level of expression of the N-*ras* gene and occurred with no change in the EC_{50} (concentration giving half-maximal response) for bombesin-stimulated inositol phosphate generation and no change in the K_d for or number of bombesin receptors on the surface of the T15 cell (Wakelam *et al.*, 1986). Further study of this system has demonstrated that bombesin stimulates the generation of $[^3H]Ins(1,4,5)P_3$ in a dose-dependent manner in $[^3H]$inositol-labelled T15+ cells, but that an increase in the second messenger is not detectable in the T15− cells (Lloyd *et al.*, 1989). In addition to stimulating an increase in $Ins(1,4,5)P_3$ generation, bombesin stimulates a much greater increase in the release of Ca^{2+} from intracellular stores in the transformed compared with the untransformed cell. Single-cell Ca^{2+} measurements demonstrated that this increase was indeed due to each individual cell within a population responding to the agonist to a greater magnitude, rather than the *ras* gene product causing an increase in the number of responsive cells (Lloyd *et al.*, 1989).

In cells transformed by activating mutations of either N- or Ha-*ras*, we detected an increase in the basal, agonist-independent rate of inositol phosphate generation (Hancock *et al.*, 1988). In the pLHT5.11 cell, the Lys-61 mutation of N-*ras* is under the control of the MMTV–LTR promotor. In this cell line, an increase in the basal rate of inositol phosphate generation was observed after a 6 h exposure to dexamethasone, in parallel experiments it was shown immunologically that it is at this time that an increase in $p21^{N\text{-}ras}$ is detectable.

To eliminate the possibility that the permanently transfected NIH-3T3 cell lines were demonstrating increases in inositol phospholipid metabolism not normally observed when cells are transformed by *ras* genes, transient transfections were performed on COS-1 cells. Transfection with the Val-12 mutant of Ha-*ras* resulted in a 1.5-fold increase in the basal rate of inositol phosphate generation (Hancock *et al.*, 1988). This increase is an underestimate of the real effect of the *ras* gene product upon inositol phospholipid metabolism, since immunofluorescence analysis demonstrated that the gene product was detectable in only approximately 20% of the cells in each culture dish. This system was also used to examine the effects of further mutations of the *ras* gene products upon inositol phospholipid metabolism. Replacement of the Cys-186 residue by a Ser group in Val-12 Ha-*ras* abolished the ability of the protein to be correctly post-translationally modified and thus be membrane bound, such cytosolic proteins cannot transform established fibroblasts (Willumsen *et al.*, 1984). COS cells transiently expressing Val-12, Ser-186 Ha-*ras* did not exhibit an amplification of the basal rate of inositol phosphate generation (Hancock *et al.*, 1988). The Asp-38–Ala-38 point mutation within the putative effector domain abolishes the transforming ability of mutant Ha-*ras* (Sigal *et al.*, 1986). COS cells expressing Val-12, Ala-38 Ha-*ras* demonstrate a basal rate of inositol phosphate generation identical to that in untransfected cells (Fig. 3). The

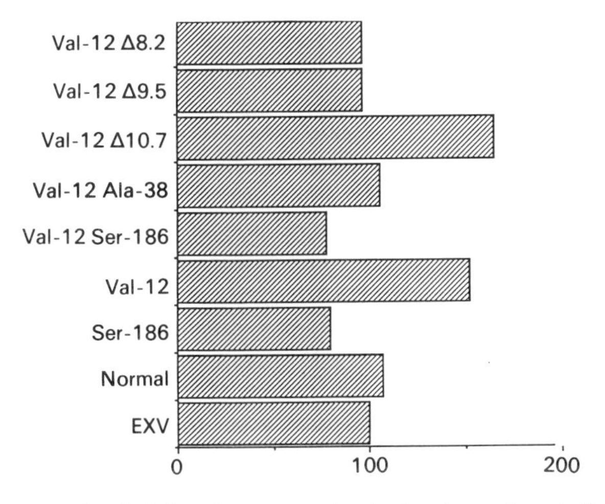

Fig. 3. *Effects of transient transfections of COS-1 cells with ras genes upon basal rates of inositol phosphate generation*

Normal Ha-*ras* and a range of mutants were transfected into COS-1 cells and the effect upon the basal rate of inositol phosphate generation determined after 60 h as described in Hancock *et al.* (1988).

correlation between the ability of a *ras* gene to transform cells and the amplification of inositol phospholipid turnover is further demonstrated by experiments where increasing amounts of the *C*-terminal region of Val-12 Ha-*ras* were deleted while leaving the five terminal amino acids intact. Three such constructs were made with only the smallest deletion (Δ 177–184) maintaining transforming ability in foci assays. Significantly, it was only this construct which was still able to amplify inositol phosphate generation in a transient transfection of COS cells (Hancock *et al.*, 1988).

While the experiments described above support the proposal that the $p21^{ras}$ proteins can function in a G_p-like manner, evidence has been presented to the contrary. In several reports a reduction rather than an amplification in inositol phospholipid breakdown in response to agonists such as platelet-derived growth factor (PDGF) (e.g. Parries *et al.*, 1987) has been observed in cells transformed by an over-expression of *ras* genes. Additionally, we have been unable to detect an amplification in bombesin-stimulated inositol phosphate generation in other transfections of NIH-3T3 cells with $p21^{N\text{-}ras}$ (Lloyd *et al.*, 1989). Consequently, we have examined a range of transfections of NIH-3T3 cells for effects upon agonist-stimulated inositol phosphate generation.

Stimulation of control, quiescent NIH-3T3 cells with prostaglandin $F_{2\alpha}$ ($PGF_{2\alpha}$) induces about a 12-fold increase in inositol phosphates; however, examination of cells transformed by the over-expression of N-, Ha-, or Ki-*ras* genes demonstrated only a 2–3-fold increase in inositol phosphate generation in response to a maximal prostaglandin concentration. This reduction was observed despite there being no change in receptor number or affinity, or in the apparent EC_{50} of the $PGF_{2\alpha}$ response (Black & Wakelam, 1990). This apparently *ras*-induced desensitization of agonist-stimulated inositol phospholipid breakdown is, however, a proliferation- rather than a transformation-related phenomenon. When normal NIH-3T3 cells grown to decreasing densities were examined, $PGF_{2\alpha}$-stimulated inositol phosphate generation was found to be decreased. At lower densities, the NIH-3T3 cells are not contact inhibited and are thus in logarithmic growth, thus in this state they are more akin to the *ras*-transformed cells which do not contact inhibit nor quiesce. Thus in dividing NIH-3T3 cells, whether transformed by a *ras* gene or not, $PGF_{2\alpha}$-stimulated in inositol phosphate generation is desensitized.

Agonist-stimulated inositol phospholipid breakdown is a short-lived event. C-kinase-mediated desensitization of agonist-stimulated inositol phosphate generation has been demonstrated in a range of cell types (see, e.g. Brown *et al.*, 1987). An increase in DAG level and protein kinase C activity has been demonstrated in *ras*-transformed fibroblasts (Lacal *et al.*, 1987; Hagag *et al.*, 1987). Indeed, in some of the *ras*-transformed cell lines in which $PGF_{2\alpha}$-stimulated inositol phosphate generation is desensitized, there is an increase in both total and membrane-bound protein kinase C activity (Fig. 4). Thus the desensitization of $PGF_{2\alpha}$-stimulated inositol phospholipid breakdown in *ras*-transformed cells is probably a reflection of activated protein kinase C and may reflect a prior increase in stimulated inositol phosphate generation. This hypothesis is supported by the ability of 12-*O*-tetradecanoyl phorbol 13-acetate (TPA) treatment of both control and *ras*-transformed cells to further desensitize the $PGF_{2\alpha}$-stimulated response (Black & Wakelam, 1990). Further support is provided by the demonstration that the desensitization is induced heterologously rather than homologously (Black & Wakelam, 1990). Thus a *ras*-induced amplification of any agonist's stimulation

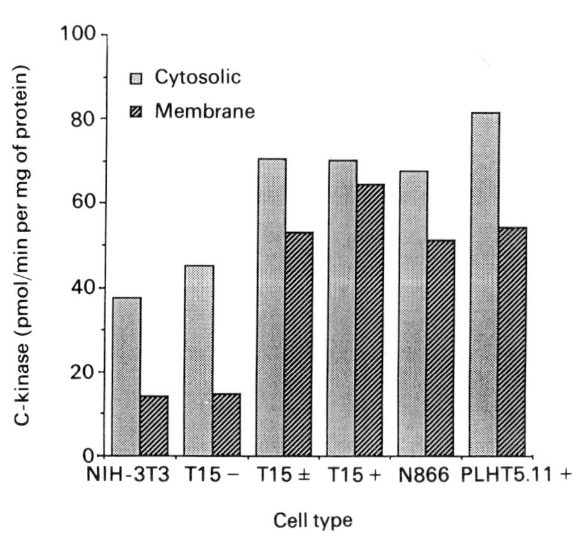

Fig. 4. *C-kinase activity in control and ras-transformed cells*

Membrane and cytosolic fractions were prepared from homogenates of cells by centrifugation. Protein kinase C activity was determined after partial purification by electrophoresis.

of inositol phospholipid breakdown will desensitize the subsequent response to other agonists. However, if there is a reduction in agonist-stimulated inositol phospholipid breakdown, it is necessary to explain the source of the elevated DAG level in *ras*-transformed cells.

Phosphatidylcholine Turnover

In a range of cell types in response to a range of agonists stimulated breakdown of phosphatidylcholine (PtdCho) has been observed in addition to that of the inositol phospholipids (see Pelech & Vance, 1989 for a review). In bombesin-stimulated Swiss 3T3 cells, the increase in Ins(1,4,5)P_3 mass peaks after 5 s, but declines rapidly reaching basal level within 1 min; continued exposure of the cells to the agonist does not induce a further increase in the mass of the second messenger (Cook *et al.*, 1990). Mass measurements of DAG, in the same cells, however, demonstrate a biphasic increase in concentration. The first increase mirrors the changes in Ins(1,4,5)P_3 while the second rise in DAG mass is observed in the absence of an increase in Ins(1,4,5)P_3 (Cook *et al.*, 1990). This second phase of DAG generation is sustained for at least 60 min and will thus maintain protein kinase C in an active state as is necessary for cell proliferation. Examination of the turnover of other cellular phospholipids has demonstrated that the secondary rise in DAG appears to be derived from PtdCho breakdown (Cook & Wakelam, 1989). Surprisingly, analysis of the water-soluble products of PtdCho hydrolysis has demonstrated that choline is generated before choline phosphate, thus the reaction is catalysed by a phospholipase D- rather than a phospholipase C-type enzyme (Cook & Wakelam, 1989). The lipid product of this reaction is phosphatidic acid which implies that the regulation of phosphatidate phosphohydrolase, the enzyme which generates DAG from phosphatidate, is a key point in the control of cell proliferation.

In *ras*-transformed cells, Lacal *et al.* (1987) demonstrated an increase in the basal level of choline phosphate, with no change in the resting levels of choline or glycerophosphocholine (GroPCho) and proposed that PtdCho-specific phospholipase C activity is increased by *ras* transformation. However, Macara (1989) demonstrated a 2-fold increase in the activity of choline kinase in *ras*-transformed cells and suggested that there is no amplification of PtdCho breakdown in the transformed cell. To address this question, we determined the resting levels of choline, choline phosphate and glycerophosphocholine in equilibrium [^{3}H]choline-labelled control and *ras*-transformed cells; in addition, we determined the basal rates of generation of each metabolite. As is clearly shown in Fig. 5, there is no change in the resting level of choline between the normal and the *ras*-transformed cells; however, there is an approximate doubling in the level of choline phosphate and a halving in the level of glycerophosphate when the *ras*-transformed cells are compared with the control cells.

Table 1 shows that the rate of glycerophosphocholine production is greater than 2-fold higher in a Ha-*ras* transformed cell compared with the control NIH-3T3 cell. This would suggest an increase in phospholipase A and lyso-

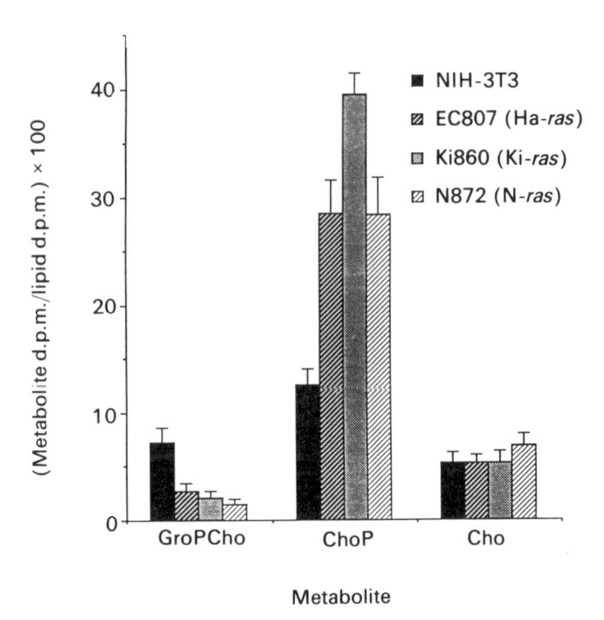

Fig. 5. *Basal levels of choline metabolites*

The basal levels of each metabolite and lipid were determined in cells labelled to equilibrium with [^{3}H]choline.

phospholipase activity in the transformed cell. Bar-Sagi & Feramisco (1986) demonstrated that microinjection of *ras* into cells resulted in an activation of phospholipase A_2; however, this may be due to the increase in intracellular free Ca^{2+} concentration or protein kinase C activity in the transformed cells. The rate of choline generation is 1.3-fold higher in the *ras*-transformed than in the control cells (Table 1); however, there is no apparent rate of choline phosphate generation (Table 1), suggesting that its removal rate equals its generation. While the basal level of choline is unchanged between the control and transformed cells (Fig. 5), its rate of production is increased, suggesting an increased rate in its metabolism. As is shown in Fig. 6, this is indeed the case—the activity of choline kinase is increased approximately 2-fold in all the *ras*-transformed cells examined with no change in the K_m of the enzyme.

Table 1. *Basal rates of generation of glycerophosphocholine (GroPCho), choline phosphate (ChoP) and choline (Cho) in control and Ha-ras (EC807) transformed NIH-3T3 cells*

Cell	Basal rate of generation [(d.p.m. in metabolite × 100/d.p.m. in lipids)/min]		
	GroPCho	ChoP	Cho
NIH-3T3	0.027 ± 0.01	0	0.036 ± 0.01
EC807	0.061 ± 0.01	0	0.047 ± 0.01
Ratio EC807 : NIH-3T3	2.23	—	1.31

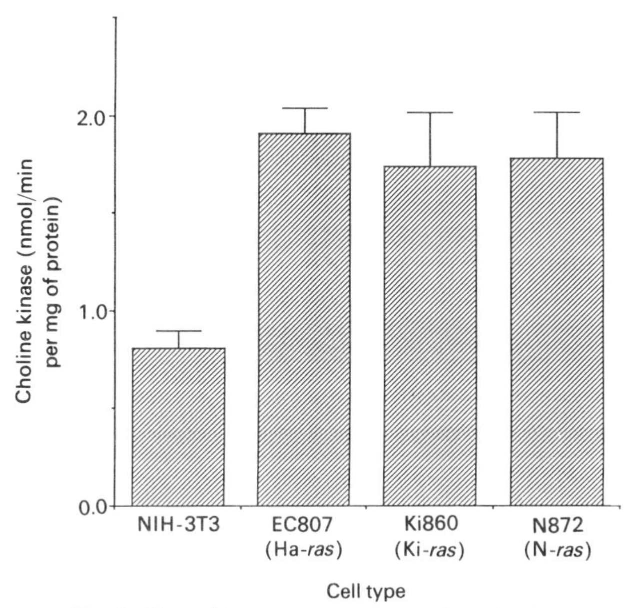

Fig. 6. *Choline kinase activity in ras-transformed cells*

Choline kinase activity was determined as described by Macara (1989) on cytosolic extracts prepared by centrifugation of cell homogenates.

In the *ras*-transformed cell there is an increase in the secretion of autocrine growth factors such as transforming growth factor α and PDGF-like molecules (Owen & Ostrowski, 1987); thus during determinations of basal turnover of PtdCho (Fig. 5) autocrines will be present and may account for the amplification observed in the *ras*-transformed cell. Therefore, the effects of agonist stimulation on PtdCho breakdown were examined in control and Ha-*ras* transformed (EC807) cells. Fig. 7 shows that the stimulated increase in choline generation in response to bradykinin and $PGF_{2\alpha}$ is amplified in EC807 cells compared with the control; no differences were observed between the cell types in the response to bombesin, PDGF or calf serum. No effect of any agonist was found upon the levels of choline phosphate or glycerophosphocholine, thus the agonists are stimulating a phospholipase D-catalysed hydrolysis of PtdCho, rather than a phospholipase C type as initially proposed by Lacal *et al.* (1987). The increase in the response to bradykinin may be due to the reported increase in the number of receptors for this peptide on Ha-*ras*-transformed cells (Parries *et al.*, 1987); however, there is no increase in the number of $PGF_{2\alpha}$ receptors on the EC807 cell (Black & Wakelam, 1990). $PGF_{2\alpha}$-stimulated choline generation showed the same agonist dose dependency as its stimulation of inositol phosphate generation with an EC_{50} of 0.17 μM; there was no difference in the EC_{50} values between the control and the transformed cells.

In a range of cell types, the C-kinase-activating phorbol ester, TPA, has been shown to stimulate phospholipase D activity (see e.g. Cook & Wakelam, 1989). Consequently, the increase in choline generation in the *ras*-transformed

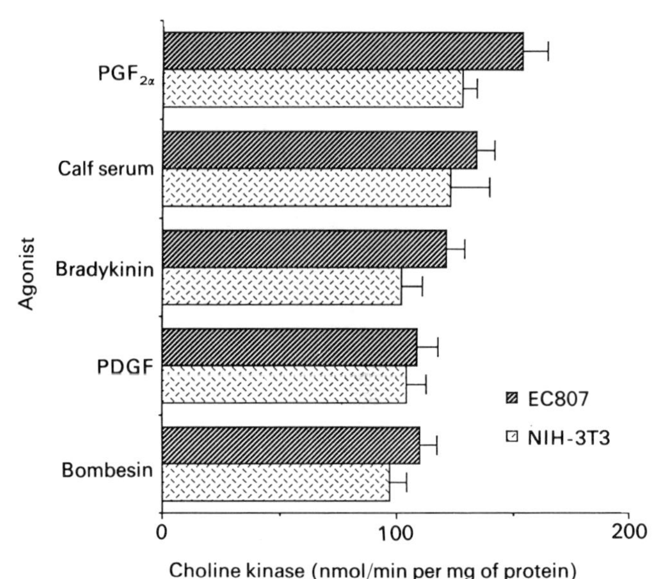

Fig. 7. *Agonist-stimulated choline generation*

Cells grown on 24-well plates and labelled to equilibrium with [^{3}H]choline were stimulated with 2.5 μM-bombesin, 1.32 μg of PDGF/ml, 3.2 μM-bradykinin, 4% (v/v) calf serum or 2.1 μM-$PGF_{2\alpha}$ for 30 min and choline generation was determined as described by Cook & Wakelam (1989).

cells may be due to an increase in C-kinase activity rather than being due to the *ras* gene product. However, this is unlikely to be the case, since stimulation of choline generation by TPA in NIH-3T3 and EC807 cells shows an identical maximum response and EC_{50}. This result would also suggest that the amplification is not due to an increase in the amount of phospholipase D in the *ras*-transformed cells, though a possible increase in the level of a non-C-kinase-sensitive form of phospholipase D cannot be ruled out at this stage.

Thus in *ras*-transformed cells there appears to be an amplification in stimulated PtdCho turnover. A similar increase has also been detected by Price *et al.* (1989), who scrape-loaded mutant p21$^{Ha\text{-}ras}$ into Swiss 3T3 cells. In agreement with the results in Fig. 5, these authors also detected an increase in apparent phospholipase A activity. Whether stimulated PtdCho activity is the primary effect of p21^{ras} is unclear. A range of studies have suggested that phospholipase D activity is stimulated secondary to prior inositol phospholipid hydrolysis (see Pelech & Vance, 1989), indeed it is possible that choline generation is increased concomitant with the desensitization of inositol phosphate generation (Cook & Wakelam, 1989; Cook *et al.*, 1990). Thus the amplification of $PGF_{2\alpha}$-stimulated choline generation and the desensitization of $PGF_{2\alpha}$-stimulated inositol phosphate generation may both be a reflection of a p21^{ras}-induced amplification of PtdIns(4,5)P_2 hydrolysis.

Conclusions

Despite intensive research it remains unclear which, if any, signal transduction pathway is the primary site of action of the *ras* gene products. It is possible to conclude that activation of protein kinase C is induced by $p21^{ras}$; however, the mechanism of this activation remains contentious. Several groups have now demonstrated that the DAG required to activate C-kinase may be derived from PtdCho breakdown. Considerable evidence, derived from a range of different cell types, suggests that stimulated PtdCho hydrolysis is secondary to prior activation of PtdIns(4,5)P_2 breakdown, a rapidly desensitized process. Thus, the p21 proteins may be acting in the first case to amplify inositol lipid hydrolysis and then, perhaps as a consequence of a C-kinase-catalysed phosphorylation of some component, to activate phospholipase D-catalysed PtdCho breakdown. This would provide for a sustained activation of protein kinase C and an amplification in the short lived Ca^{2+} signal, both of which appear to be necessary for growth factor-stimulated cell proliferation (Milligan & Wakelam, 1989).

Work from this laboratory is supported by grants from the C.R.C. and the M.R.C. (U.K.).

References

Ballester, R., Michaeli, T., Ferguson, K., Xu, H. P., McCormick, F. & Wigler, M. (1989) *Cell (Cambridge, Mass.)* **59**, 681–686
Barbacid, M. (1987) *Annu. Rev. Biochem.* **56**, 779–827
Bar-Sagi, D. & Feramisco, S. R. (1986) *Science* **233**, 1061–1068
Berridge, M. J. (1987) *Biochim. Biophys. Acta* **907**, 33–45
Berridge, M. J. & Irvine, R. F. (1989) *Nature (London)* **341**, 197–205
Black, F. M. & Wakelam, M. J. O. (1990) *Biochem. J.* **267**, 809–813
Bos, J. L. (1988) *Mut. Res.* **195**, 255–271
Boyer, J. L., Helper, J. R. & Harden, T. K. (1989) *Trends Pharmacol. Sci.* **10**, 360–365
Brown, K. D., Blakeley, D. M., Hamon, M. H., Laurie, S. M. & Corps, A. N. (1987) *Biochem. J.* **245**, 631–639
Cook, S. J. & Wakelam, M. J. O. (1989) *Biochem. J.* **263**, 581–587
Cook, S. J., Palmer, S., Plevin, R. & Wakelam, M. J. O. (1990) *Biochem. J.* **265**, 617–620
Davies, S. A., Houslay, M. D. & Wakelam, M. J. O. (1989) *Biochim. Biophys. Acta* **1013**, 173–179
Der, C. J., Finkel, T. & Cooper, G. M. (1986) *Cell (Cambridge, Mass.)* **44**, 167–176
DeVos, A. M., Tong, L., Milburn, M. V., Matias, P. M., Jancarik, J., Noguchi, S., Nishimura, S., Miura, K., Ohtsuka, E. & Kim, S. H. (1988) *Science* **239**, 888–893
Feursten, J., Goody, R. S. & Wittinghofer, A. (1987) *J. Biol. Chem.* **262**, 8455–8460
Fleischman, L. F., Chawala, S. B. & Cantley, L. (1986) *Science* **231**, 407–410
Franks, D. J., Whitfield, J. F. & Durkin, J. P. (1987) *Biochem. Biophys. Res. Commun.* **147**, 596–601
Hagag, N., Lacal, J. C., Graber, M., Aaronson, S. & Viola, M. V. (1987) *Mol. Cell. Biol.* **7**, 1984–1988
Hall, A. & Self, A. (1986) *J. Biol. Chem.* **261**, 10963–10965
Hancock, J. F., Marshall, C. J., McKay, I. A., Gardner, S., Houslay, M. D., Hall, A. & Wakelam, M. J. O. (1988) *Oncogene* **3**, 187–193
Hancock, J. F., Magee, A. I., Childs, J. E. & Marshall, C. J. (1989) *Cell (Cambridge, Mass.)* **57**, 1167–1177
Kamata, T., Sullivan, N. F. & Wooten, M. W. (1987) *Oncogene* **1**, 37–46
Lacal, J. C., Moscat, J. & Aaronson, S. (1987) *Nature (London)* **330**, 269–272
Lloyd, A. C., Davies, S. A., Crossley, I., Whitaker, M., Houslay, M. D., Hall, A., Marshall, C. J. & Wakelam, M. J. O. (1989) *Biochem. J.* **260**, 813–819
Macara, I. G. (1989) *Mol. Cell. Biol.* **9**, 325–328
Milligan, G. & Wakelam, M. J. O. (1989) *Prog. Growth Factor Res.* **1**, 161–177
Milligan, G., Davies, S. A., Houslay, M. D. & Wakelam, M. J. O. (1989) *Oncogene* **4**, 659–663
Owen, R. D. & Ostrowski, M. C. (1987) *Mol. Cell. Biol.* **7**, 2512–2520
Parries, G., Hoebel, R. & Racker, I. (1987) *Proc. Natl. Acad. Sci. U.S.A.* **84**, 2648–2652
Pelech, S. L. & Vance, D. E. (1989) *Trends Biochem. Sci.* **14**, 28–30
Price, B. D., Morris, J. D. H., Marshall, C. J. & Hall, A. (1989) *J. Biol. Chem.* **264**, 16638–16643

Seeburg, P. H., Colby, W. W., Capon, D. J., Goeddel, D. V. & Levinson, A. D. (1984) *Nature (London)* **312**, 71–75

Sibley, D. R., Benovic, J., Caron, M. G. & Lefkowitz, R. J. (1987) *Cell (Cambridge, Mass.)* **48**, 913–922

Sigal, I. S., Gibbs, J. B., D'Aonzo, J. S., Temeles, G. L., Wolanski, B. S., Socher, S. H. & Scolnick, E. M. (1986) *Proc. Natl. Acad. Sci. U.S.A.* **83**, 952–956

Tagliaferri, P., Kutsaros, D., Clair, T., Neckers, L., Robins, R. K. & Cho-Chung, Y. S. (1988) *J. Biol. Chem.* **263**, 409–416

Toda, T., Uno, I., Ishikawa, T., Powers, S., Kataoka, T., Broek, D., Cameran, S., Broach, J., Matsumoto, K. & Wigler, M. (1985) *Cell (Cambridge, Mass.)* **40**, 27–36

Trahey, M. & McCormick, F. (1987) *Science* **238**, 542–545

Vogel, U., Dixon, R. A. F., Schaber, M. D., Diehl, R. E., Marshall, M. S., Scolnick, G. M., Sigal, I. S. & Gibbs, J. B. (1988) *Nature (London)* **335**, 90–93

Wakelam, M. J. O., Davies, S. A., Houslay, M. D., McKay, I., Marshall, C. J. & Hall, A. (1986) *Nature (London)* **323**, 173–176

Willumsen, B. M., Norris, K., Papageorge, A. G., Hubbert, N. L. & Lowy, D. R. (1984) *EMBO J.* **3**, 2581–2585

Biochem. Soc. Symp. **56**, 117–136
Printed in Great Britain

Chronic Ethanol-Induced Heterologous Desensitization is Mediated by Changes in Adenosine Transport

ADRIENNE S. GORDON*†‡, LAURA NAGY*†, DARIA MOCHLY-ROSEN*†‡ and IVAN DIAMOND*†‡§

**Ernest Gallo Clinic and Research Center, San Francisco General Hospital, San Francisco, CA 94110, U.S.A. and Departments of †Neurology, §Pediatrics and ‡Pharmacology, University of California, San Francisco, CA 94143, U.S.A.*

Synopsis

Chronic exposure of cultured cell lines to ethanol results in a heterologous desensitization of receptors coupled to adenylate cyclase via G_s, the stimulatory guanine nucleotide regulatory protein. This heterologous desensitization is accompanied by a decrease in α_s, the GTP-binding subunit of G_s. Ethanol-induced accumulation of extracellular adenosine is required for the development of heterologous desensitization. To determine the mechanism underlying the ethanol-dependent increase in extracellular adenosine, we investigated the effects of ethanol on the nucleoside transporter responsible for the bidirectional transport of adenosine into and out of the cell. We found that ethanol specifically and non-competitively inhibited nucleoside uptake. Inhibition of adenosine uptake was primarily due to decreased influx via the nucleoside transporter. Thus, ethanol-induced increases in extracellular adenosine appear to be due to inhibition of adenosine influx. After chronic exposure to ethanol, cells became tolerant to the acute effects of ethanol, i.e. ethanol no longer inhibited uptake and, consequently, no longer increased extracellular adenosine concentration. Taken together with our previous studies, these results suggest that acute ethanol inhibition of adenosine influx leads to an increase in extracellular adenosine which in turn activates adenosine A_2 receptors to increase cyclic AMP levels, leading to desensitization of receptor-dependent cyclic AMP signal transduction after chronic exposure to ethanol.

We next determined whether the same effects of ethanol also occur in alcoholics. We isolated lymphocytes from alcoholics and non-alcoholics and found that alcoholics had a 75% decrease in basal and adenosine receptor-stimulated cyclic AMP production compared with non-alcoholics. To determine whether these differences were due to exposure to ethanol *in vivo* or to a possible genetic difference between alcoholics and non-alcoholics, we grew lymphocytes in culture in the absence of ethanol. Adenosine receptor-stimulated cyclic AMP levels were higher in alcoholics than non-alcoholics. Moreover, cultured cells from alcoholics were more sensitive to the effects of chronic alcohol on cyclic AMP signal transduction than cells from non-alcoholics. Our results suggest that the cyclic AMP signal transduction system may reflect a genetic alteration in alcoholics and that studies in cultured lymphocytes may allow us to identify individuals at risk of developing alcoholism.

Introduction

The neurological and behavioural complications of alcoholism are major problems in medicine and in society. Alcoholics become tolerant to ethanol and exhibit physical dependence after chronic alcohol abuse. The acute and chronic effects of ethanol in many cell types, and particularly in the nervous system, appear to be due to its interaction with cell membranes [1–4]. Chronic exposure to ethanol induces changes in the cell, most likely in the cell membrane, which lead to dependence on ethanol and tolerance to the effects of ethanol. However, the actual membrane changes which account for tolerance and dependence to ethanol are not understood. Acute and chronic ethanol-induced changes in membrane order and membrane constituents (e.g. cholesterol/phospholipid ratios [5–8] and fatty acids [9–11]) have been reported, but experiments using different animals, different strains of the same animals, or the same strains in different laboratories, have yielded conflicting results [12]. This may be due in part to genetic factors, the use of disrupted tissue preparations, and the mode of ethanol administration. A most serious problem derives from the heterogeneity of brain. Changes restricted to selected brain regions or cell types, or to specific membrane components, may be difficult to detect in crude brain preparations. Finally, it is nearly impossible to distinguish primary biochemical effects of ethanol from secondary responses related to systemic, metabolic and hormonal influences in such systems. To overcome these methodological limitations, we have developed a model cell culture preparation using the neuroblastoma × glioma hybrid cell line, NG108-15, and S49 lymphoma cells in defined medium to study the responses of intact cells to acute and chronic exposure to ethanol under precisely controlled conditions [13].

We have chosen to study the cyclic AMP signal transduction system, since it requires sequential activation of three membrane proteins and would be expected to be sensitive to physical and biochemical changes in the membrane. Moreover, since cyclic AMP levels regulate many different cellular functions, changes in cyclic AMP might be responsible for the pleiotropic effects of ethanol. Since recent evidence suggests that adenosine may play a role in mediating the effects of ethanol in the brain [14,15], we studied the effects of acute and chronic exposure to ethanol on adenosine receptor-dependent cyclic AMP accumulation in cultured cell lines. We have also extended our studies to freshly isolated and cultured lymphocytes from alcoholics and non-alcoholics.

Cultured Cell Lines

Adenosine receptor-stimulated cyclic AMP signal transduction system

N^6-Phenyisopropyladenosine (PIA) is an adenosine agonist that stimulates cyclic AMP production in a dose-dependent manner. Fig. 1 shows that PIA-

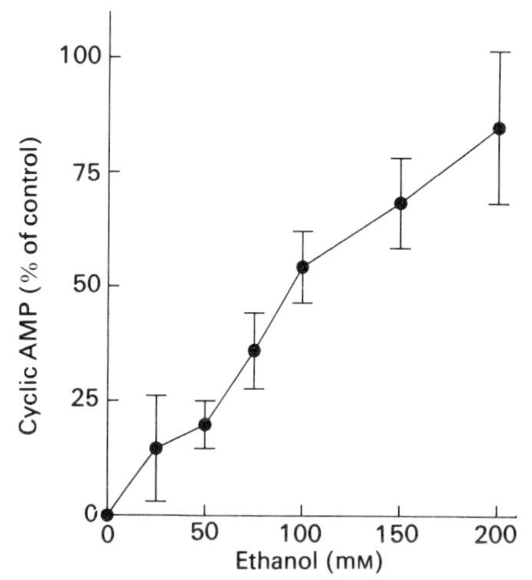

Fig. 1. *Acute ethanol stimulation of cyclic AMP levels in NG108-15 cells in the presence of 0.1 mM-PIA*

Cells were grown in defined medium depleted of oleic acid, insulin and transferrin, preincubated for 60 min with the phosphodiesterase inhibitor, Ro20-1724, and incubated with PIA in the presence or absence of ethanol for 30 min.

stimulated cyclic AMP accumulation at a maximum PIA concentration is further increased in NG108-15 cells with increasing concentrations of ethanol added during the PIA incubation period [16]. Concentrations of ethanol as low as 50 mM cause a 20–40% increase in PIA-stimulated cyclic AMP levels.

Since acute ethanol exposure increases PIA stimulation of cyclic AMP production, cellular adaptation to chronic ethanol exposure might be expressed by a reduction in adenosine receptor-activated adenylate cyclase activity. To test this possibility, NG108-15 cells were grown in medium containing 200 mM-ethanol for 48 h [16]. Fig. 2 shows that there was a 35% reduction in PIA-stimulated cyclic AMP levels after 48 h in ethanol when measured in the absence of ethanol. If this decrease in receptor-activated cyclic AMP accumulation is a compensatory response related to chronic ethanol exposure, i.e. a form of dependence on ethanol, then only in the presence of ethanol should the chronically treated cells have the same cyclic AMP levels as cells never exposed to ethanol. Therefore, cells grown in medium for 2 days in 200 mM-ethanol were washed and rechallenged with 200 mM-ethanol. Fig. 2 shows that chronically treated cells still respond to acute ethanol exposure by increasing PIA-stimulated cyclic AMP levels. Most importantly, in the presence of ethanol, the chronically treated cells show PIA-stimulated cyclic AMP levels which are the same as those in control cells.

If this response is related to cellular dependence, then the adenosine receptor-stimulated cyclic AMP levels should return to normal after withdrawal of ethanol from chronically treated cells. Fig. 2 shows that 48 h after removal of ethanol, PIA-stimulated cyclic AMP levels in recovered cells are similar to those in control cells never exposed to ethanol. In addition, acute stimulation in cells withdrawn from ethanol is the same as in control cells.

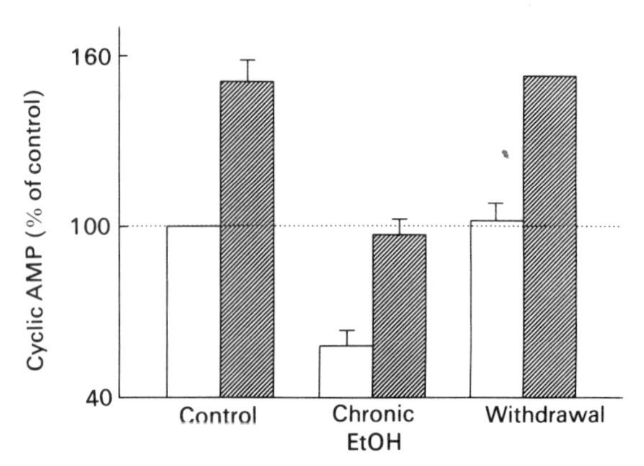

Fig. 2. *Ethanol-induced changes in adenosine receptor-dependent cyclic AMP levels in NG108-15 cells*

Cells were grown without (control) or with 200 mM-ethanol (chronic EtOH) for 48 h ($n = 9$). All cells were treated with PIA to activate adenosine receptors in the absence (□) or presence (▨) of 200 mM-ethanol. In a separate experiment, cells were grown with 100 mM-ethanol for 48 h and kept without ethanol for an additional 48 h (withdrawal). Cyclic AMP levels were assayed in the absence ($n = 5$) or presence ($n = 2$) of 100 mM-ethanol. Each bar represents the mean ± S.E.M.

Heterologous desensitization

Using the NG108-15 cell line, which has both adenosine and prostaglandin E_1 (PGE_1) receptors that activate adenylyl cyclase, we could determine whether ethanol affects more than one receptor in the same cell [17]. We found that PGE_1 receptor-stimulated cyclic AMP production was decreased by 25.6% in this cell line after a 48 h exposure to 100 mM-ethanol, similar to the decrease in adenosine receptor-mediated cyclic AMP production (Table 1). This result suggests that chronic ethanol treatment causes heterologous desensitization of receptors that stimulate adenylyl cyclase activity.

α_s is decreased in cells treated chronically with ethanol

Production of cyclic AMP requires at least three membrane components: receptors for hormones and neurotransmitters, a GTP-binding protein, G_s, and the enzyme that synthesizes cyclic AMP, adenylyl cyclase. Alterations in G_s could account for the reduced stimulation of cyclic AMP production by both adenosine and PGE_1. Therefore, we investigated whether chronic ethanol treatment alters the amount and/or function of α_s, the subunit of G_s that mediates receptor stimulation of adenylyl cyclase.

To assay α_s function, S49 lymphoma *cyc*$^-$ cells were reconstituted with the α_s-subunit in a Lubrol extract of membranes from control and from ethanol-treated NG108-15 cells. Cells that are *cyc*$^-$ lack α_s and therefore cannot activate adenylylcyclase via receptor stimulation unless supplemented with exogenous α_s. The amount of cyclic AMP produced in this system in the presence of the β-receptor agonist isoprenaline and a non-hydrolysable GTP analogue, guanylimidodiphosphate thus reflects the amount and/or the ability of the added α_s to couple receptor activation to adenylyl cyclase. Consistent with the

Table 1 *NG108-15 cells were treated with 100 mM-ethanol for 48 h*

Adenosine receptor-dependent cyclic AMP levels were determined in intact control and ethanol-treated cells as described in Fig. 8. Cyclic AMP assays, and Western and Northern blot analyses were carried out on the same cells. The numbers in parentheses indicates the number of independent experiments on cell cultures grown at different times. The amount of α_s mRNA was also determined by slot blot analysis using 0.625, 1.25, 2.5 and 5.0 μg of total RNA. The results are from four separate RNA preparations analysed on four Northern blots and one slot blot. Western, Northern and slot blots were analysed by scanning autoradiograms with a Hoeffer GS300 densitometer connected to a Hewlett–Packard 3390A integrator. Each band was analysed by scanning four separate segments and the areas from each scan were added to compensate for variations in band-width across the lane. Results are expressed as percentage decrease (mean ± S.E.M.) of ethanol-treated cells compared with controls.

	Decrease (%)
Cyclic AMP production	
Adenosine receptor-dependent	27.0 ± 6.0 (4)
PGE_1 receptor-dependent	25.6 ± 5.7 (3)
α_s function (*cyc*$^-$ reconstitution)	29.0 ± 1.0 (3)
α_s protein (Western blot analysis)	38.5 ± 5.5 (4)
α_s mRNA (Northern blot analysis)	30.0 ± 2.0 (4)

decrease in adenosine- and PGE_1-mediated cyclic AMP production in intact cells, chronic ethanol treatment caused a 29% reduction in the ability of membrane extracts to complement the α_s deficiency of *cyc*$^-$ membranes (Fig. 3 and Table 1). This result was not due to any alteration in α_s extraction efficiency after chronic ethanol treatment (data not shown).

Decreased amounts of α_s protein in the ethanol-treated cells could be responsible for the decreased cyclic AMP production in the reconstitution assay with *cyc*$^-$ membranes. Using anti-α_s antibodies in a Western blot analysis, we measured membrane-bound α_s in control and ethanol-treated cells. Fig. 4 shows that chronic exposure to ethanol results in a decrease in α_s protein; densitometric analysis of the autoradiograms revealed the decrease was 38.5% (Table 1). The fact that the reductions in adenosine- and PGE_1-stimulated cyclic AMP production, in the functional reconstitution assay, and in α_s protein (Table 1), were comparable, suggests that the reduction in the amount of α_s protein is sufficient to account for the decrease in receptor-stimulated cyclic AMP production. The amount of α_i, the GTP-binding subunit of the inhibitory G-protein, is not altered in these cells by chronic ethanol exposure [18]. It appears, therefore, that the absolute amount of α_s limits cyclic AMP production in NG108-15 cells.

To investigate the mechanism underlying the ethanol-induced reduction in α_s protein, we compared the amount of α_s-specific mRNA in ethanol-treated and control cells. Northern blot analysis of total cellular RNA showed a 30% reduction in α_s mRNA in the ethanol-treated cells (Fig. 5*a* and *b* and Table 1). In contrast, β-tubulin mRNA was slightly increased in the same experiments

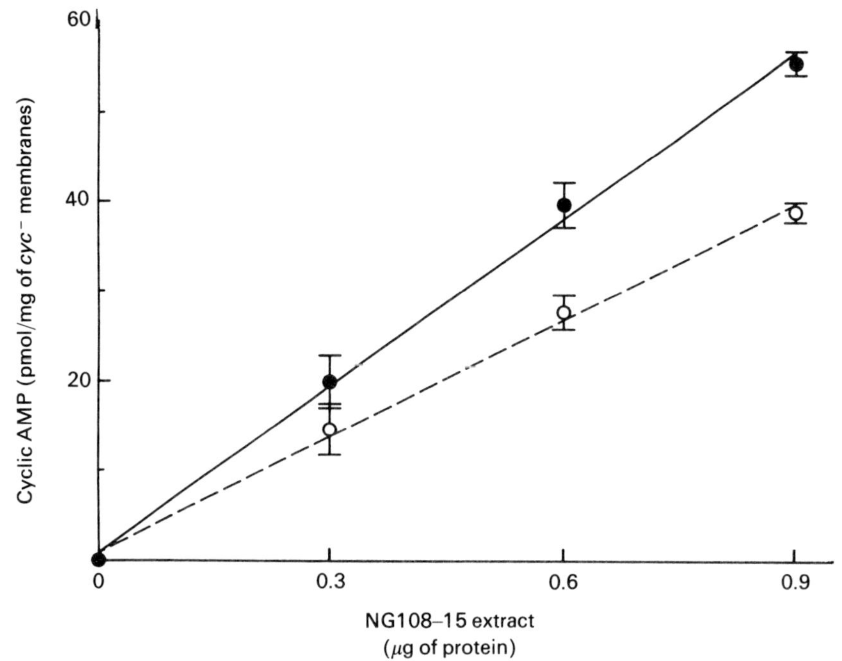

Fig. 3. *Activity of α_s in NG108-15 membranes measured by reconstitution of α_s-deficient cyc$^-$ membranes*

Increasing amounts of Lubrol PX extracts from NG108-15 membranes isolated from control (●) and ethanol-treated cells (○) were incubated with *cyc*$^-$ membranes and receptor-stimulated adenylyl cyclase activity was measured. Results are the means ± S.E.M. of three separate experiments carried out in triplicate.

(Fig. 5*c* and *d*). Thus, the ethanol-induced reduction in mRNA coding for α_s is unlikely to be due to global changes in cellular mRNA. It remains to be determined whether decreased mRNA coding for α_s is due to a reduction in transcription and/or the stability of α_s mRNA.

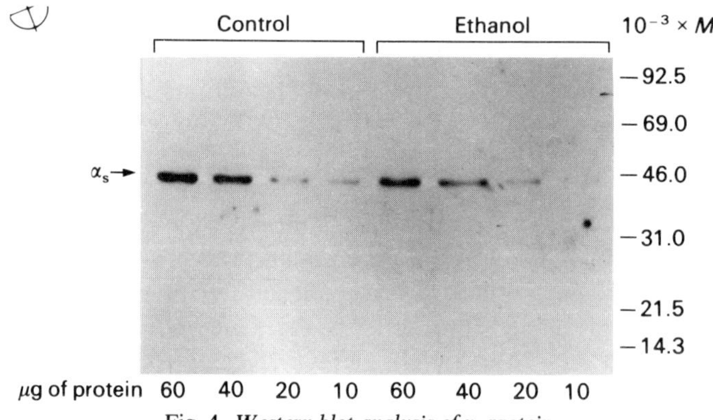

Fig. 4. *Western blot analysis of α_s protein*

Equal amounts of membrane protein from control and ethanol-treated cells were electrophoresed on SDS/polyacrylamide gels, electrotransferred to nitrocellulose paper, and then probed with rabbit anti-α_s antiserum (serum A-572, a generous gift from Drs Susanne M. Mumby & Alfred G. Gilman).

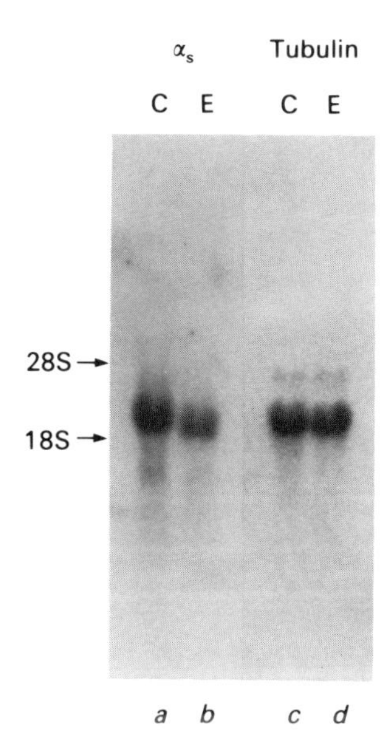

Fig. 5. *Northern blot analysis of α_s and β-tubulin mRNA*

Total RNA was isolated from control (*a,c*) and ethanol-treated (*b,d*) cells. RNA was electrophoresed on formaldehyde–agarose gels and probed with radioactive rat α_s (*a,b*) and human β-tubulin (*c,d*) cDNA probes. The cDNA probe for rat α_s (*C*-terminal two-thirds of the open-reading frame) was obtained from Dr Randall Reed. The cDNA probe for human β-tubulin (entire coding region) was obtained from Dr Larry Kedes.

Extracellular adenosine is required for ethanol-induced heterologous desensitization

We next carried out a series of experiments to determine the mechanism responsible for the heterologous desensitization of receptor-dependent cyclic AMP levels observed after chronic exposure to ethanol [19]. Since heterologous desensitization in other systems is characterized by an initial increase in cyclic AMP levels [20], we first determined whether acute exposure to ethanol caused an increase in cyclic AMP levels even in the absence of exogenously added agonist. When NG108-15 cells were incubated with 100 mM-ethanol for 10 min, there was a 60% increase in intracellular cyclic AMP levels (Fig. 6). Since ethanol does not activate adenylyl cyclase directly, but stimulates only receptor-dependent cyclic AMP production [21], we next explored the possibility that acute ethanol increases the extracellular concentration of a stimulatory agonist. Neural cells [22], lymphocytes [23,24] and other cell types [25] release adenosine, and adenosine can cause both homologous and heterologous desensitization [26,27]. Moreover, adenosine has been implicated in the central nervous system effects of ethanol [14,15]. Therefore, adenosine concentrations in the media of control and ethanol-treated cells were measured using h.p.l.c. There was a significant increase in the concentration of extracellular adenosine after NG108-15 cells were incubated with 200 mM-ethanol (Fig. 7). Within 10 min, adenosine concentrations reached 37 ± 1.2 nM/5×10^6 cells in

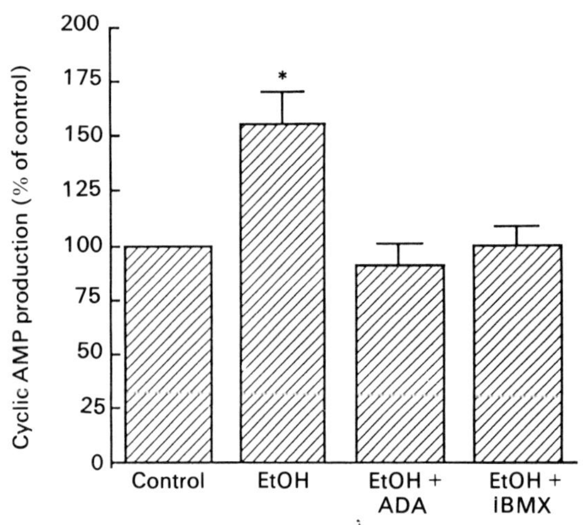

Fig. 6. *Acute effect of ethanol on endogenous cyclic AMP levels in NG108-15 neuroblastoma × glioma cells in the absence of added agonist*

Cells were incubated with or without 100 mM-ethanol (EtOH) in the presence or absence of 1 unit of ADA/ml or 10 mM-IBMX. Cyclic AMP levels in the absence of ethanol were 18.4 ± 2.4 pmol/10^6 cells (16 experiments). Values are expressed as the percentage of cyclic AMP levels in cells not exposed to ethanol. Bars represent means ± S.E.M. ($n = 4$–9). * $P < 0.002$ compared with cells not treated with ethanol.

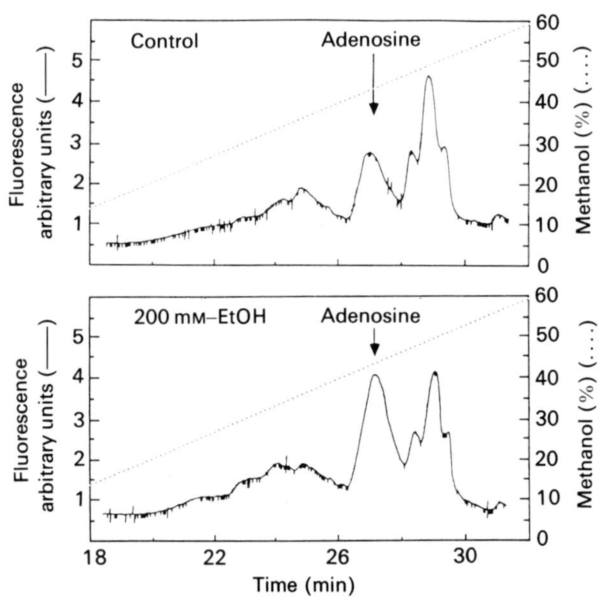

Fig. 7. *Acute effect of ethanol on extracellular adenosine concentration in NG108-15 cells*

Cells were incubated with or without 200 mM-ethanol (EtOH) for 10 min. The extracellular adenosine concentration was determined by h.p.l.c. Representative chromatographs of control and ethanol-treated cells are shown.

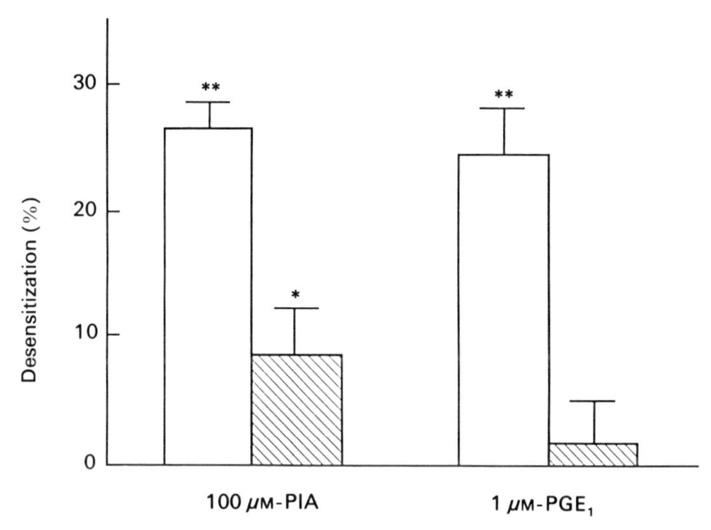

Fig. 8. *Chronic effects of ethanol on desensitization of receptor-stimulated cyclic AMP production in NG108-15 cells*

Cells were maintained for 48 h with or without 100 mM-ethanol in the presence (▧) or absence (□) of 1 unit of ADA/ml. Cells were washed and incubated with 100 μM-PIA or 1 μM-PGE_1 for 30 min in the absence of ethanol. Desensitization is expressed as the percentage decrease in ethanol-treated cells compared with cells never exposed to ethanol; cells treated with ethanol and ADA are compared with cells treated with ADA alone. *$P < 0.025$, **$P < 0.001$ compared with cells not treated with ethanol (*$P < 0.025$ ** $P < 0.001$, Student's *t*-test).

ethanol-treated cells while control cultures had 18.2 ± 3.7 nM-adenosine ($n = 4$, $P < 0.005$).

Adenosine modulates the production of cyclic AMP via A_1 and A_2 adenosine receptors [28]. NG108-15 cells have only the A_2 adenosine receptor (K. C. Collier & A. S. Gordon, unpublished work) which is positively coupled to adenylylcyclase [28]. Therefore, the increase in intracellular cyclic AMP levels produced by acute ethanol (Fig. 6) could be due to an ethanol-induced increase in extracellular adenosine concentration. If this were the case, then degradation of extracellular adenosine should prevent stimulation of cyclic AMP production by ethanol. We used adenosine deaminase (ADA) to deaminate adenosine to inosine, a nucleoside with low affinity for the adenosine receptor [28]. When NG108-15 cells were incubated with ADA, stimulation of cyclic AMP production by ethanol was completely abolished (Fig. 6). Moreover, treatment of the cells with an adenosine receptor antagonist, 3-isobutyl-1-methylxanthine (IBMX), also completely blocked ethanol-induced increases in cyclic AMP levels (Fig. 6). These data suggest that acute exposure to ethanol caused an increase in extracellular adenosine. This extracellular adenosine then activated the A_2 receptor to stimulate cyclic AMP production.

In contrast to acute stimulation of cyclic AMP levels by ethanol, chronic exposure to ethanol causes a decrease or desensitization of adenosine receptor- and PGE_1 receptor-dependent cyclic AMP production (Figs. 2 and 8). If adenosine were responsible for ethanol-induced heterologous desensitization, ADA should prevent this response. Fig. 8 shows that when NG108-15 cells are co-incubated for 48 h with ethanol and 1 unit of ADA/ml, a concentration sufficient to block the acute increase in cyclic AMP (Fig. 6), chronic ethanol-induced desensitization of adenosine receptor-stimulated cyclic AMP levels is substantially blocked and desensitization of the PGE_1 receptor is completely prevented. These results suggest that an ethanol-induced increase in extracellular adenosine concentration is required for ethanol to produce heterologous desensitization. In other preparations, desensitization by adenosine is dependent on the concentration of agonist and time of exposure [26]. Consistent with our results, Green [22] has found, in C1300 neuroblastoma cells, that 10–20 nM of endogenously released adenosine is sufficient to desensitize receptor-dependent cyclic AMP production during an overnight culture [22]. Moreover, we find that after exposure to 70 nM-PIA for 48 h, adenosine receptor-dependent cyclic AMP levels were decreased by 26% (data not shown), a similar reduction to that induced by 100 mM-ethanol for 48 h.

The requirement for extracellular adenosine in ethanol-induced heterologous desensitization was not specific for neural cells. S49 wild-type lymphoma cells showed a significant increase in the concentration of extracellular adenosine when exposed to ethanol for 5 min or 24 h (Table 2). When S49 wild-type cells were treated with 100 mM-ethanol for 48 h, adenosine receptor- and PGE_1 receptor-stimulated cyclic AMP levels were reduced to 65% and 36% of control, respectively (Table 2). As in NG108-15 cells, addition of ADA to S49 wild-type cells prevented ethanol-induced heterologous desensitization (Table 2).

Table 2. *Ethanol-induced accumulation of extracellular adenosine and heterologous desensitization in S49 wild-type and adenosine-transport mutants (80-2A6 and 160-D4)*

Extracellular adenosine concentrations were determined by h.p.l.c. after exposure of 1.2×10^7 cells in a total volume of 20 ml to 100 mM-ethanol for 24 h and 10^7 cells in a total volume of 1 ml to 200 mM-ethanol for 5 min. S49 wild-type cells were grown in the absence (Wild-type) or presence of 1.5 units of ADA/ml (Wild-type + chronic ADA). Basal levels of cyclic AMP were 3.54 ± 0.70, 3.58 ± 0.41 and 3.43 ± 1.11 pmol of cyclic AMP/10^6 cells, and PGE_1-stimulated cyclic AMP levels were 25.9 ± 7.2, 62.0 ± 25.0 and 34.0 ± 10.2 pmol of cyclic AMP/10^6 cells for S49 wild-type, 80-2A6 and 160-D4, respectively. Cyclic AMP levels of ethanol-treated cells (100 mM, 48 h) are expressed as a percentage of cyclic AMP in cells never exposed to ethanol. Values represent means ± S.E.M. when $n > 2$, mean ± range when $n = 2$. Number of determinations (n) is indicated in parentheses. * $P < 0.001$.

	Extracellular adenosine (nM/10^7 cells)			Cyclic AMP levels (% of control)	
Cell type	No EtOH		EtOH	100 μM-PIA	1 μM-PGE_1
Wild-type					
5 min	4.4 ± 0.9 (5)	200 mM	17.6 ± 3.6 (5)	65 ± 7* (27)	36 ± 6* (6)
24 h	21.9 ± 0.3 (2)	100 mM	56.7 ± 10.2 (2)		
Wild-type + chronic ADA	ND		ND	118 ± 31 (4)	96 ± 17 (5)
80-2A6					
5 min	0 (2)	200 mM	0 (2)	115 ± 11 (6)	82 ± 7 (4)
24 h	0 (3)	100 mM	0 (3)		
160-D4	ND		ND	94 ± 9 (4)	98 ± 7 (4)

If accumulation of extracellular adenosine is required for ethanol-induced heterologous desensitization, then cells which do not release adenosine should not desensitize after chronic exposure to ethanol. Adenosine uptake and release are mediated via a single bidirectional transporter which can be blocked by agents such as dipyridamole [29–31]. However, we have found that concentrations of dipyridamole necessary to inhibit transport are toxic to S49 cells over a 48 h period (L. E. Nagy, unpublished work). Therefore, the nucleoside transport-deficient mutants 80-2A6 and 160-D4 of the S49 lymphoma cell line [31,32] were utilized to determine if adenosine transport is required for ethanol-induced heterologous desensitization. When the 80-2A6 mutant cell line was treated with 200 mM-ethanol for 5 min or 100 mM-ethanol for 24 h, extracellular adenosine was not detectable (Table 2). There was also no desensitization of adenosine receptor- or PGE_1 receptor-stimulated cyclic AMP levels when the adenosine transport-deficient cells were exposed to 100 mM-ethanol for 48 h (Table 2). Thus, the adenosine transporter is required for ethanol-induced heterologous desensitization of receptor-dependent cyclic AMP production in S49 cells.

The lack of desensitization after chronic exposure to ethanol in the adenosine transport-deficient cells was not due to altered stimulation of cyclic AMP by agonist. Incubation of S49 wild-type cells with maximally effective concentrations of PIA or PGE_1 increased cyclic AMP levels 1.92 ± 0.41-fold ($n = 18$) and 13.5 ± 3.5-fold ($n = 6$) over basal, respectively. Similar results were obtained in the adenosine transport-deficient cells (see the legend to Table 2), indicating that coupling of adenosine receptor and PGE_1 receptor to adenylyl cyclase is normal in the mutant cells. Although the adenosine transporter is a non-specific nucleoside carrier [31,33,34], ethanol-induced desensitization does not appear to be due to other nucleosides, since they have very low affinities for the adenosine receptor and do not stimulate cyclic AMP production [28]. Moreover, exposure of S49 wild-type cells to inosine for 48 h did not desensitize the adenosine receptor (data not shown). Taken together, our results suggest that accumulation of extracellular adenosine is required for ethanol-induced heterologous desensitization.

Ethanol increases extracellular adenosine by inhibiting adenosine uptake via the nucleoside transporter

Adenosine has a variety of regulatory functions including modulation of neurotransmitter release [35], regulation of immune function [35,36], inhibition of lipolysis [37] and inhibition of platelet aggregation [38]. The majority of these actions are mediated by adenosine binding to specific receptors on the cell membrane. Inhibitors of adenosine uptake potentiate cellular responses to extracellular adenosine [36,39–41], suggesting that the physiological effects of adenosine are terminated by re-uptake of adenosine into the cell [42].

In mammalian cells, adenosine is transported by facilitated diffusion via the nucleoside transporter. This transporter has a broad specificity for nucleosides and mediates both influx and efflux [29–31,33,34]. After influx through the nucleoside transporter, adenosine is subject to intracellular metabolism

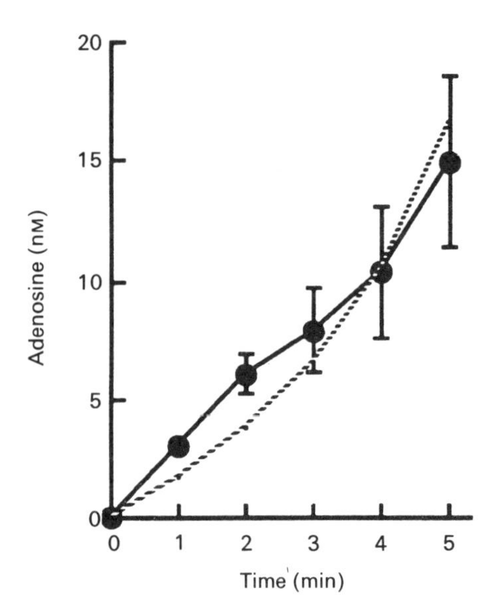

Fig. 9. *Time-dependent increase in extracellular adenosine concentration in the presence of 200 mM-ethanol*

Aliquots of S49 cells in suspension (2×10^7 cells/ml) were incubated without ethanol for 5 min. Ethanol was then added to a final concentration of 200 mM at 5, 6, 7, 8, or 9 min. Control cells were carried out through the entire 10 min incubation without ethanol. Extracellular adenosine concentrations were measured by h.p.l.c. The *Y*-axis represents the increase in extracellular adenosine in the supernatant of cells treated with ethanol above that found in the absence of ethanol (4.4 ± 0.9 nM). The dotted line represents the predicted increase in extracellular adenosine which would result from a 35% reduction in adenosine uptake. This line was calculated by increasing the 4.4 nM-adenosine observed in the absence of ethanol by 35% over each min of incubation in ethanol. Values represent means ± S.E.M., $n = 5$.

[33,34]. Phosphorylation of adenosine, which traps adenosine inside the cell, or deamination to inosine are the primary forms of metabolism [33,34]. Adenosine can also condense with homocysteine to form *S*-adenosylhomocysteine (SAH) [33,34]. Thus, ethanol-induced increases in extracellular adenosine could result from decreased adenosine influx and metabolism and/or increased adenosine efflux from the cell.

Extracellular adenosine

The concentration of extracellular adenosine increased with time when S49 cells were incubated in the absence of ethanol. However, addition of 200 mM-ethanol for 1–5 min increased extracellular adenosine concentrations above that found in cells not exposed to ethanol. In 5 min, there was a 4.5-fold increase in extracellular adenosine concentrations in S49 cells treated with ethanol compared with cells incubated without ethanol (Fig. 9, Table 2) [43].

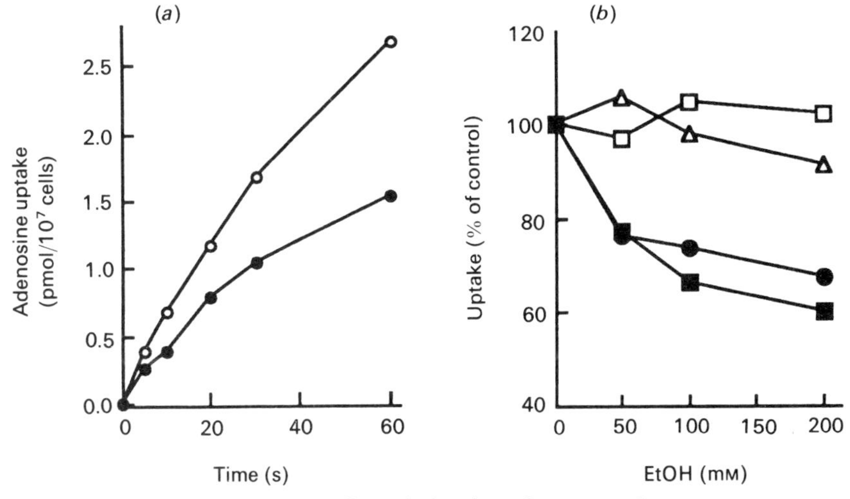

Fig. 10. *Acute effects of ethanol on adenosine uptake*

(*a*) Time course of [^{3}H]adenosine uptake in S49 cells in the presence and absence of ethanol. After preincubation for 4 min in 0 or 200 mM-ethanol, uptake of 0.3 μM-[^{3}H]adenosine was measured in the absence (○) and presence (●) of 200 mM-ethanol by rapid centrifugation through oil. Non-specific uptake was determined in the presence of 10 μM-dipyridamole and subtracted from total uptake to determine specific uptake. Non-specific uptake averaged 0.03 pmol/10^7 cells at 10 s. Values are from a representative experiment done in triplicate, replicated six times. Mean values for ethanol-treated cells from all six experiments are significantly lower than controls ($P < 0.02$). (*b*) Uptake of nucleosides, leucine, and deoxyglucose in S49 cells in the presence of varying concentrations of ethanol. Uptake of 0.3 μM-[^{3}H]adenosine (●), 0.3 μM-[^{3}H]uridine (■), 50 μM-[^{3}H]-isoleucine (□) and 50 μM-[^{3}H]deoxyglucose (△) was measured at 90 s in the presence of 50–200 mM-ethanol and compared with cells incubated in the absence of ethanol. Points represent mean values from three to six experiments done in triplicate. For greater clarity, S.E.M. values (from 6 to 12% of mean value) are not shown.

Adenosine uptake

Ethanol-induced increases in extracellular adenosine could be due to decreased adenosine uptake. The time course of adenosine uptake in S49 cells is shown in Fig. 10(*a*). Uptake was nearly linear for 60 s and inhibited by 200 mM-ethanol at all time points measured; 30–40% inhibition was detected as early as 5 s and was the same at 1 min. Thirty-five per cent inhibition of adenosine uptake at 1 min completely accounts for the rate of accumulation of extracellular adenosine in S49 cells treated with ethanol (Fig. 9).

Inhibition of adenosine uptake increased with increasing concentrations of ethanol from 50 to 200 mM (Fig. 10*b*). An Eadie–Hofstee kinetic analysis showed that acute ethanol was a non-competitive inhibitor of uptake in S49 cells. $V_{max.}$ for uptake was decreased by 50% in ethanol-treated cells, with only a small change in affinity [43]. Ethanol also inhibited the uptake of uridine, another nucleoside transported by the nucleoside transporter [33,34], but had no effect on isoleucine and deoxyglucose uptake (Fig. 10). Inhibition of adenosine uptake by alcohols increased with increasing chain length (data not shown).

Adenosine influx via the nucleoside transporter

The measurement of adenosine uptake includes influx through the nucleoside transporter and subsequent intracellular metabolism. To investigate ethanol inhibition of influx directly, we used a non-metabolizable analogue of adenosine, 5′-deoxy[5-^{3}H]adenosine. This analogue has been used by others as a non-metabolizable substrate in several cells of lymphoid origin [44,45]. Our initial experiments showed that 5′-deoxy[5-^{3}H]adenosine was metabolized by S49 cells (data not shown). However, h.p.l.c. analysis indicated that 5′-deoxy[5-^{3}H]adenosine was not metabolized by the human lymphoblastoid cell line, CCRF-CEM. Therefore, these cells were used for further study. Ethanol inhibition of [^{3}H]adenosine uptake in CCRF-CEM cells was similar to that in S49 cells (Fig. 11*a*). Moreover, influx of 5′-deoxy[5-^{3}H]adenosine was also inhibited by 200 mM-ethanol (Fig. 11*b*). These results suggest that ethanol inhibition of adenosine uptake in CCRF-CEM cells is due primarily to decreased influx.

Adenosine metabolism

The activities of ADA, adenosine kinase and SAH hydrolase *in vitro* were not altered by adding 200 mM-ethanol directly to isolated cytosol preparations, or after acute exposure of intact cells to 200 mM-ethanol for 10 min or chronic exposure to 100 mM-ethanol for 48 h [43].

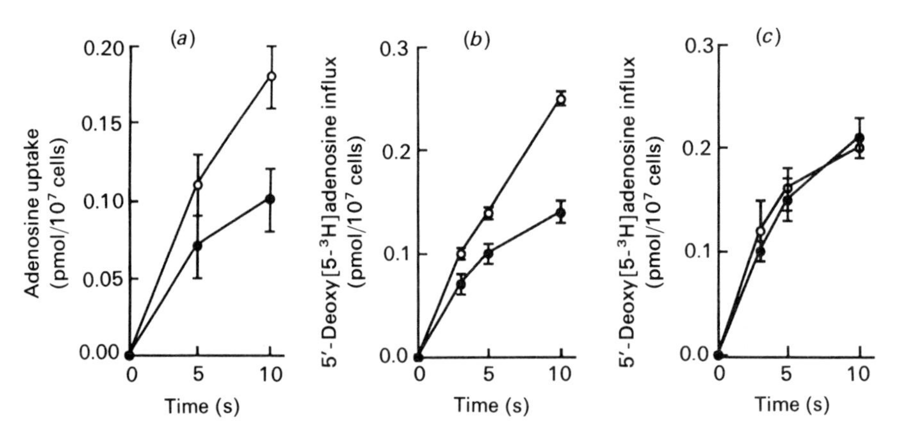

Fig. 11. *Time course for ethanol inhibition of adenosine uptake and 5′-deoxy [5-^{3}H]adenosine influx in control and chronically treated cells*

CCRF-CEM cells, grown in the absence (*a*,*b*) or presence (*c*) of 100 mM-ethanol for 48 h, were preincubated without (○) or with (●) 200 mM-ethanol for 4 min. Uptake of [^{3}H]adenosine (*a*) and influx of 5′-deoxy[5-^{3}H]adenosine (*b*,*c*) were determined over 3–10 s. Values represent mean ± S.E.M. (n = 4–5). In (*a*) and (*b*) values in ethanol-treated cells are significantly less than uptake in the absence of ethanol at all time points ($P < 0.01$).

5′-Deoxyadenosine efflux

Ethanol-induced increases in extracellular adenosine might also be due to an increase in adenosine efflux. CCRF-CEM cells were preloaded with 5′-deoxy[5-^{3}H]adenosine and efflux measured in the presence or absence of ethanol. Since efflux was too rapid to measure at 22°C, incubations were carried out at 12°C. Dipyridamole completely blocked 5′-deoxy[5-^{3}H]adenosine efflux under these conditions, indicating that efflux occurred via the nucleoside transporter [43]. Release of labelled 5′-deoxy[5-^{3}H]adenosine did not differ between control and ethanol-treated cells [43].

Chronic effects of ethanol on adenosine uptake and extracellular adenosine concentration

We examined the chronic effects of ethanol on the adenosine transport system. After exposure to 200 mM-ethanol for 48 h, acute ethanol no longer inhibited adenosine uptake in S49 cells (Fig. 12*a*). Influx of the non-metabolizable adenosine analogue, 5′-deoxy[5-^{3}H]adenosine, also became insensitive to acute ethanol inhibition in chronically treated CCRF-CEM cells (Fig. 11*c*). If ethanol-induced increases in extracellular adenosine result from inhibition of uptake, then ethanol should not increase extracellular adenosine in chronically treated cells. Fig. 12 shows that, after exposure to 100 mM-ethanol for 48 h, acute ethanol no longer increased extracellular adenosine concentrations in S49 cells.

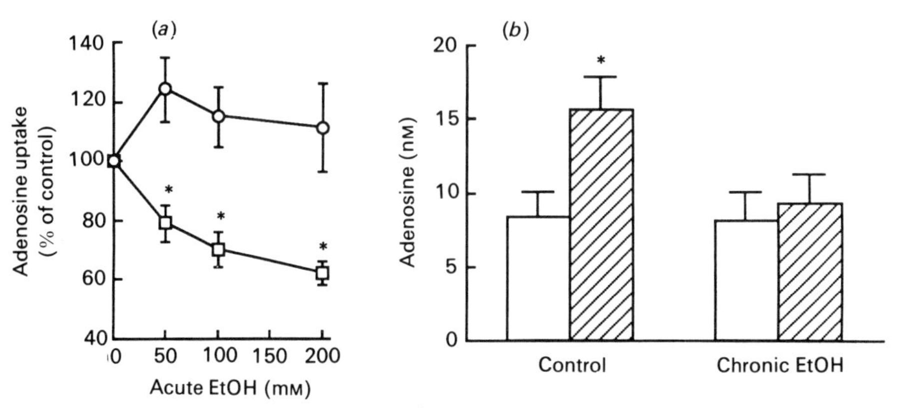

Fig. 12. *Chronic effects of ethanol on adenosine uptake and extracellular adenosine concentration*

(*a*) Chronic effects of ethanol on [^{3}H]adenosine uptake. S49 cells were grown in the absence (□) or presence (○) of 100 mM-ethanol for 48 h. Adenosine uptake was measured at 90 s in the presence of 50–200 mM-ethanol and compared with cells incubated in the absence of ethanol. Values represent means ± S.E.M., $n = 6$–11. * $P < 0.01$ compared with uptake in the absence of ethanol. (*b*) Ethanol-induced extracellular adenosine accumulation in S49 cells after chronic ethanol exposure. Cells were grown in the absence (Control) or presence (Chronic EtOH) of 100 mM-ethanol for 48 h. Media were removed and S49 cells (1×10^7 cells/ml) incubated in the absence (□) or presence (▨) of 100 mM-ethanol for 10 min (end panel). Extracellular adenosine concentrations were determined by h.p.l.c. Values represent means ± S.E.M. $n = 10$–14. * $P < 0.003$ compared with cells not treated with acute ethanol.

These data suggest that acute exposure to clinically relevant concentrations of ethanol appears to decrease adenosine uptake by inhibiting influx. In contrast, we found no effect of ethanol on efflux. After prolonged exposure to ethanol, S49 and CCRF-CEM cells became tolerant to the acute inhibitory effects of ethanol on adenosine uptake. Ethanol no longer inhibited uptake of adenosine. Inhibition of adenosine uptake by ethanol results in an increase in extracellular adenosine. Since ethanol-induced increases in extracellular adenosine are required for chronic ethanol-induced heterologous desensitization of receptors coupled to adenylyl cyclase via G_s, it is possible that ethanol-induced changes in the adenosine transport system could account for many of the acute and adaptive responses to ethanol.

Cyclic AMP Signal Transduction in Alcoholics

Desensitization of adenosine receptor-dependent cyclic AMP production in alcoholics

To determine whether the increase in receptor-dependent cyclic AMP levels we observed in cultured cell lines after chronic exposure to ethanol might be of pathophysiological significance in chronic alcoholism, we tested whether cells taken from actively drinking alcoholics would also show heterologous desensitization [46]. Since human lymphocytes have the same A_2 adenosine receptors as NG108-15 and S49 cells, we could test directly whether alcoholics have altered cyclic AMP signal transduction. We undertook a controlled study of basal and adenosine receptor-stimulated cyclic AMP levels in lymphocytes of chronic alcoholics, normal subjects, and patients with non-alcoholic liver disease.

Under standard conditions of assay for cyclic AMP, unstimulated (basal) and PIA-stimulated levels of cyclic AMP in normal human lymphocyte preparations ($n = 31$) were 8.97 ± 0.88 (S.E.M.) and 14.85 ± 1.37 pmol/10^6 cells, respectively. There was no correlation of basal or stimulated cyclic AMP levels with age or sex, but day-to-day variation was noted in some subjects.

Alcoholics showed highly significant depression of basal and PIA-stimulated cyclic AMP levels in intact lymphocytes when compared to normal subjects (Fig. 13) or patients with liver disease. Mean basal cyclic AMP levels (pmol/10^6 cells) in controls, alcoholics and patients with liver disease were 9.55 ± 1.65, 2.30 ± 0.34 ($P = 0.0004$) and 8.33 ± 1.29 ($P = 0.0005$), respectively. Mean PIA-stimulated cyclic AMP levels were 15.81 ± 2.52, 3.72 ± 0.53 ($P = 0.0002$) and 14.04 ± 1.93 ($P = 0.0005$), respectively. The difference between alcoholics and controls was also striking when the effect of 80 mM-ethanol on PIA-stimulated cyclic AMP levels was compared (Fig. 13). The percentage response to ethanol was decreased markedly in lymphocytes from alcoholics compared with cells from normal subjects. Ethanol increased PIA-stimulated cyclic AMP levels by 17% in lymphocytes from alcoholics compared with 71% in normal individuals ($P = 0.002$). The mean cyclic AMP level in PIA-stimulated lymphocytes from alcoholics after addition of ethanol was 4.33 ± 0.96 pmol/10^6 cells, while the value for normal controls was

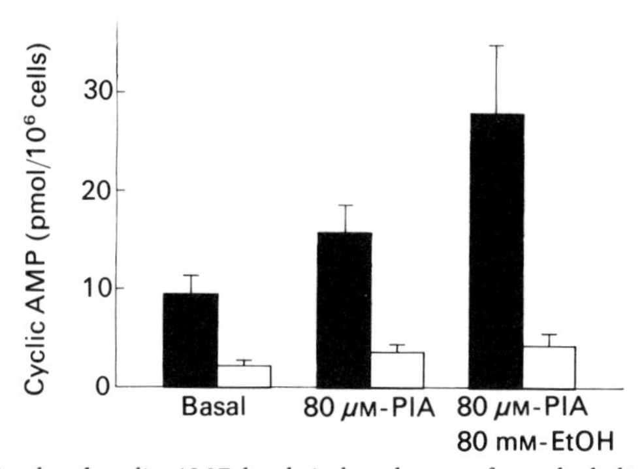

Fig. 13. *Basal and stimulated cyclic AMP levels in lymphocytes from alcoholics and control subjects*
PIA is an adenosine receptor agonist. Each bar represents the mean ± S.E.M. (n = 10 for basal and PIA; n = 9 for PIA plus ethanol). □, Alcoholics; ■, control subjects.

28.0 ± 5.22. These results suggest that basal and receptor-stimulated cyclic AMP levels in lymphocytes may be valuable as a sensitive test for distinguishing between actively drinking alcoholic and non-alcoholic human populations.

Our results suggest that lymphocytes from alcoholics are distinguishable from lymphocytes of non-alcoholic subjects. Cells from alcoholics exhibit reduced basal cyclic AMP levels, reduced PIA-stimulated cyclic AMP levels and increased resistance to ethanol stimulation of adenosine receptor-dependent cyclic AMP accumulation. These changes do not appear to be due to liver disease *per se*, and it is unlikely that malnutrition can account for these results. The alcoholic patients appeared well-nourished clinically, they had normal weight for height and lacked laboratory evidence of malnutrition. Therefore, it is likely that the reduced levels of cyclic AMP in freshly isolated lymphocytes from alcoholics may reflect an acquired membrane abnormality caused by chronic alcohol abuse. These conclusions are supported by our findings that two abstinent alcoholics (2 years and 5 years since the last drink), had similar basal and adenosine receptor-stimulated cyclic AMP levels to non-alcoholics.

Genetic factors

There is also compelling evidence that genetic factors play a role in the development of alcoholism [47]. Since the cyclic AMP signal transduction system is altered by ethanol, genetic changes affecting this system might lead to a susceptibility to alcoholism. We determined that lymphocytes from alcoholics, after growth for four to six generations without ethanol, do exhibit differences in signal transduction [48]. Cells from alcoholics not only recover their responsiveness to adenosine receptor stimulation when grown in the absence of ethanol, but have a 2.8-fold greater stimulation of cyclic AMP accumulation by PIA than lymphocytes cultured from non-alcoholics (Fig. 14). This difference is not due to differences in the percentage of T and B cells [46]. There is also no difference in doubling time between alcoholics

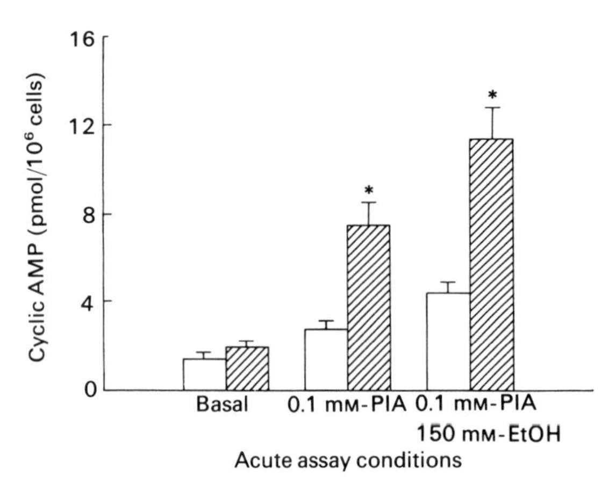

Fig. 14. *Adenosine receptor-stimulated cyclic AMP levels in cultured lymphocytes from alcoholic and non-alcoholic subjects*

Lymphocytes from alcoholic (□; $n = 7$) and non-alcoholic (▨; $n = 9$) subjects were grown for 7 days in the absence of ethanol and adenosine-stimulated cyclic AMP levels measured. * $P < 0.001$ compared with non-alcoholic subjects.

(26.9 ± 1.8 h) and non-alcoholics (23.7 ± 1.7 h). In addition to this difference in the absence of ethanol, lymphocytes from alcoholics are also more sensitive to ethanol than those from non-alcoholics. After challenge with only 100 mM-ethanol for 24 h, lymphocytes from alcoholics exhibited a 39% decrease in

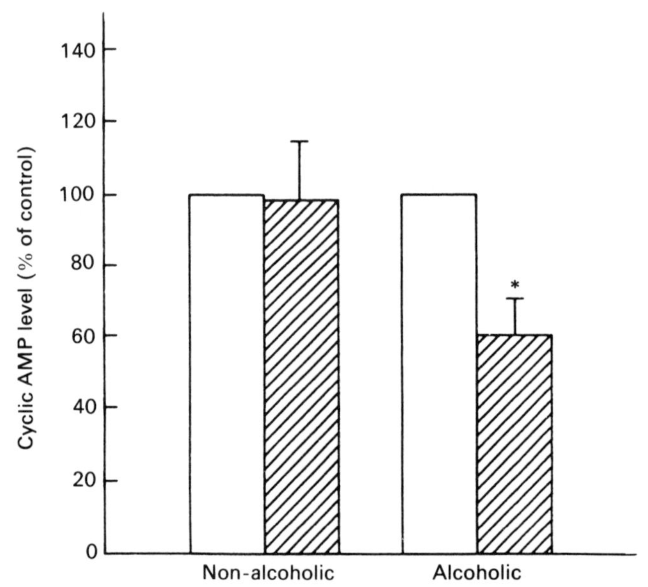

Fig. 15. *Difference in sensitivity to chronic ethanol exposure between cells from alcoholic and non-alcoholic subjects*

Lymphocytes from alcoholic ($n = 7$) and non-alcoholic ($n = 9$) subjects were grown in culture for 7 days; 100 mM-ethanol was added to half of the culture flasks for the last 24 h (▨) and PIA-stimulated cyclic AMP levels were determined. Values represent cyclic AMP levels in ethanol-treated cells as a percentage of cyclic AMP levels in the same cells grown without ethanol (□). * $P < 0.003$ compared with non-alcoholic subjects.

adenosine receptor-stimulated cyclic AMP levels, while lymphocytes from non-alcoholics showed no change under these conditions (Fig. 15).

Goldstein & Goldstein [49] have proposed that drug dependency develops as a cell or organism makes homoeostatic adjustments which compensate for the primary effect of a drug. In the case of ethanol, acute stimulation of cyclic AMP levels by ethanol appears to be countered by a chronic decrease in receptor-stimulated cyclic AMP levels [16,50–55]. In the presence of acute ethanol and PIA, cells from alcoholics exhibit a 2.8-fold increase in cyclic AMP levels compared with cells from non-alcoholics (Fig. 14). It is therefore possible that these higher amounts of cyclic AMP could cause the cells of alcoholics to adapt at lower concentrations of ethanol (Fig. 15).

In summary, our data suggest that suppressed adenosine receptor-stimulated cyclic AMP levels in freshly isolated lymphocytes from alcoholics represents both an acquired abnormality caused by long-term exposure to ethanol in the blood and a genetically determined alteration in cellular function. Differences in adenosine receptor-dependent cyclic AMP signal transduction observed in lymphocytes from alcoholics could be due to changes in any component of the signal transduction pathway (receptors, G-protein, adenylate cyclase cyclic AMP-dependent protein kinase, adenosine transporter). By growing lymphocytes in culture without ethanol, it will be possible to identify which component of the adenosine receptor-mediated cyclic AMP signal transduction system is abnormal in alcoholics.

References

1. Goldstein, D. B. (1983) *Pharmacology of Alcohol*, Oxford University Press, New York
2. Chin, J. H. & Goldstein, D. B. (1981) *Mol. Pharmacol.* **19**, 425–431
3. Franks, N. P. & Lieb, W. R. (1982) *Nature (London)* **300**, 487–493
4. Rottenberg, H., Waring, A. & Rubin, E. (1981) *Science* **213**, 583–584
5. Johnson, D. A., Lee, N. M., Cooke, R. & Loh, H. H. (1979) *Mol. Pharmacol.* **15**, 739–746
6. Chin, J. H., Parsons, L. M. & Goldstein, D. B. (1978) *Biochim. Biophys. Acta* **513**, 358–363
7. Parsons, L. M., Gallaher, E. J. & Goldstein, D. B. (1982) *J. Pharmacol. Exp. Ther.* **223**, 472–476
8. Smith, T. L. & Gerhart, M. J. (1982) *Life Sci.* **31**, 1419–1425.
9. Waring, A. J., Rottenberg, H., Ohnishki, T. & Rubin, E. (1981) *Proc. Natl. Acad. Sci. U.S.A.* **78**, 2582–2586
10. Littleton, J. M. & John, G. R. (1977) *J. Pharm. Pharmacol.* **28**, 579–580
11. Reitz, R. C., Wang, L., Schilling, R. J., Starich, G. H., Bergstrom, J. D. & Thompson, J. A. (1982) *Prog. Lipid Res.* **20**, 209–213
12. Lyon, R. C. & Goldstein, D. B. (1983) *Mol. Pharmacol.* **23**, 86–91
13. Charness, M. E., Gordon, A. S. & Diamond, I. (1983) *Science* **222**, 1246–1248
14. Dar, M. S., Mustafa, S. J. & Wooles, W. R. (1983) *Life Sci.* **33**, 1363–1374
15. Proctor, W. R. & Dunwiddie, T. V. (1984) *Science* **226**, 519–521
16. Gordon, A. S., Collier, K. & Diamond, I. (1986) *Proc. Natl. Acad. Sci. U.S.A.* **83**, 2105–2108
17. Mochly-Rosen, D., Chang, F.-U., Cheever, L., Kim, M. & Diamond, I. (1988) *Nature (London)* **333**, 848–850
18. Charness, M. E., Querimit, L. A. & Henteleff, M. (1988) *Biochem. Biophys. Res. Commun.* **155**, 138–143
19. Nagy, L. E., Diamond, I., Collier, L., Lopez, L., Ullman, B. & Gordon, A. S. (1989) *Mol. Pharm.* **34**, 744–748
20. Sibley, D. R. & Lefkowitz, R. J. (1985) *Nature (London)* **317**, 124–129
21. Rabin, R. A. & Molinoff, P. B. (1983) *J. Pharmacol. Exp. Ther.* **227**, 551–556
22. Green, R. D. (1980) *J. Supramol. Struct.* **13**, 175–182
23. Newby, A. C. & Holmquist, C. A. (1981) *Biochem. J.* **200**, 399–403

24. Fredholm, B. B., Sandberg, G. & Ernstrom, E. (1978) *Biochem. Pharmacol.* **27**, 2675–2682
25. Stone, T. W. (ed.) (1985) *Purines: Pharmacology and Physiological Roles*, Macmillan Press Ltd, London
26. Kenimer, J. G. & Nirenberg, M. (1981) *Mol. Pharmacol.* **20**, 585–591
27. Newman, M. E. & Levitzki, A. (1983) *Biochem. Pharmacol.* **32**, 137–140
28. Daly, J. W. (1985) in *Advances in Cyclic Nucleotides and Protein Phosphorylation* (Cooer, D. M. F. & Seamon, K. B., eds.), vol. 19, pp. 29–46, Raven Press, New York
29. Taube, R. A. & Berlin, R. D. (1972) *Biochem. Biophys. Acta* **255**, 6–18
30. Oliver, J. M. & Paterson, A. R. P. (1971) *Can. J. Biochem.* **49**, 262–270
31. Cohen, A., Ullman, B. & Martin, D. W., Jr (1979) *J. Biol. Chem.* **254**, 112–116
32. Ullman, B., Kaur, K. & Watts, T. (1983) *Mol. Cell Biol.* **3**, 1187–1196
33. Plagemann, P. G. W. & Wohlhueter, R. M. (1980) *Curr. Top. Membr. Transp.* **14**, 225–330
34. Wu, P. H. & Phillis, J. W. (1984) *Neurochem. Int.* **6**, 613–632
35. Fox, I. H. & Kelley, W. N. (1978) *Annu. Rev. Biochem.* **47**, 655–686
36. Berne, R. M., Rall, T. W. & Rubio, R. (eds.) (1983) *Regulatory Function of Adenosine*, Martinus Nijhoff Publishers, The Hague
37. Schwabe, U., Eber, R. & Erbler, H. C. (1973) *Naunyn-Schmiedeberg's Arch. Pharmacol.* **276**, 133–148
38. Haslam, R. J. & Rosson, G. M. (1975) *Mol. Pharmacol.* **11**, 528–544
39. Bonnafous, J. C., Bornand, J., Favero, J. & Mani, J. C. (1981) *J. Receptor Res.* **2**, 347–366
40. Baer, H. P. (1983) *Eur. J. Pharmacol.* **89**, 185–191
41. Phillis, J. W., Edstrom, J. P., Kostopoulos, G. K. & Kirkpatrick, J. R. (1984) *Can. J. Physiol. Pharmacol.* **57**, 1289–1312
42. Jarvis, S. M. (1988) *Adenosine Receptors*, pp. 113–123, Alan R. Liss, New York
43. Nagy, L. E., Diamond, I., Casso, D. J., Franklin, C. & Gordon, A. S. (1990) *J. Biol. Chem.* **265**, 1946–1951
44. Kessel, D. (1978) *J. Biol. Chem.* **253**, 400–403
45. Plagemann, P. G. W. & Wohlhueter, R. M. (1983) *Biochem. Pharmacol.* **32**, 1433–1440
46. Diamond, I., Wrubel, B., Estrin, W. & Gordon, A. (1987) *Proc. Natl. Acad. Sci. U.S.A.* **84**, 1413–1416
47. Cloninger, R. C. (1987) *Science* **236**, 410–416
48. Nagy, L. E., Diamond, I. & Gordon, A. (1988) *Proc. Natl. Acad. Sci. U.S.A.* **85**, 6973–6976
49. Goldstein, D. B. & Goldstein, A. (1961) *Biochem. Pharmacol.* **8**, 48
50. Bode, D. C. & Molinoff, P. B. (1985) *Fed. Proc. Fed. Am. Soc. Exp. Biol.* **44**, 1239
51. Richelson, E., Stenstrom, S., Forray, C., Enloe, L. & Pfenning, M. (1986) *J. Pharmacol. Exp. Ther.* **239**, 687–692
52. Saito, T., Lee, J. M. & Tabakoff, B. (1985) *J. Neurochem.* **44**, 1037–1044
53. Valverius, P., Hoffman, P. L. & Tabakoff, B. (1987) *Mol. Pharmacol.* **32**, 217–222
54. Hoffman, P. L. & Tabakoff, B. (1986) *J. Neurochem.* **46**, 812–816
55. Saito, T., Lee, J. M., Hoffman, P. L. & Tabakoff, B. (1987) *J. Neurochem.* **48**, 1817–1822

Biochem. Soc. Symp. **56**, 137–154
Printed in Great Britain

Guanine-Nucleotide-Binding Proteins in Diabetes and Insulin-Resistant States

MARK BUSHFIELD, SUSANNE L. GRIFFITHS, DEREK STRASSHEIM, ERIC TANG, YASMIN SHAKUR, BRIAN LAVAN and MILES D. HOUSLAY

Molecular Pharmacology Group, Institute of Biochemistry, University of Glasgow, Glasgow G12 8QQ, Scotland, U.K.

Introduction

Many receptors exert actions upon target cells by controlling the production of intracellular second messengers. In a number of instances they do this by modulating the activity of a distinct protein species involved in the production of an intracellular 'second message'. Thus one finds receptors which can express either stimulatory or inhibitory functions on cellular signal transduction systems. The coupling of this large family of membrane-bound receptors to their membrane-bound signal generation systems appears for a large family of receptors to involve a specific class of coupling proteins called guanine-nucleotide-regulatory proteins (G-proteins). These are heterotrimeric species consisting of α-, β- and γ-subunits. Of these, it is the α-subunit which defines the role of the G-protein; being responsible for both binding and hydrolysing GTP, as well as determining the interaction with both the appropriate receptor and signal generator systems [1–3].

Adenylate cyclase plays a pivotal role in controlling the functioning of a wide variety of cells. Specific receptors can either stimulate or inhibit its activity. These two classes of receptors mediate their stimulatory and inhibitory actions upon adenylate cyclase through two distinct G-proteins called G_s and G_i, respectively, which appear to have identical $\beta\gamma$-subunits associated with them. Appropriate receptors determine the functioning of these G-proteins by altering their conformational state which allows them to bind GTP and dissociate to release the $\beta\gamma$ components. The free α-subunit thus adopts an activated configuration and then interacts with adenylate cyclase. In the case of $G_s\alpha$, then, activation of adenylate cyclase ensues. However, with G_i, the situation is more complex. Inhibition (Fig. 1) is probably caused by both the action of $G_i\alpha$ on adenylate cyclase and also by the action of the $\beta\gamma$ components released from G_i which, by mass action, attenuate the dissociation and hence activation of G_s. These α-subunits are unique and can be distinguished by the ability of specific bacterial toxins to catalyse their NAD^+-dependent ADP-ribosylation. Thus, cholera toxin catalyses the ADP-ribosylation and activation of G_s, while pertussis toxin ADP-ribosylates and inactivates G_i [1–4].

Direct, receptor-independent, activation of G-proteins can also be achieved in both broken membranes and in permeabilized cells by the addition of non-

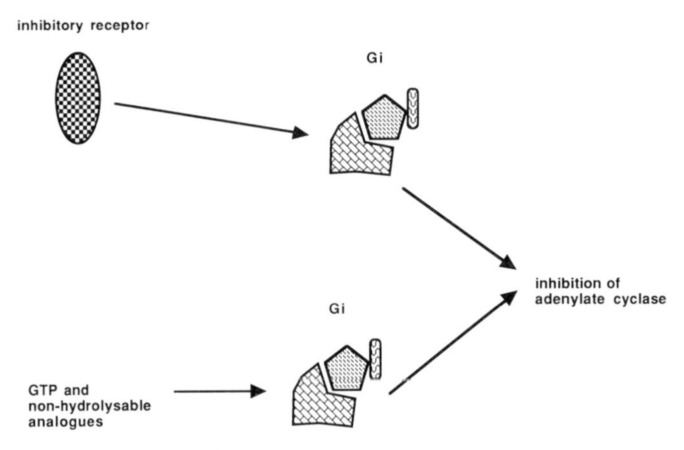

Fig. 1. *Modes of inhibition of adenylate cyclase by* G_i

'G_i' consists of α-, β- and γ-subunits. Activation can be achieved either by virtue of occupancy of an appropriate receptor or by using non-hydrolysable GTP analogues to activate 'G_i' directly and thus short circuit the need for receptor occupancy. Under physiological conditions, GTP concentrations are sufficiently high enough in both adipocytes and hepatocytes for G_i to be partially activated and exert a 'tonic' inhibitory function. Inhibition can, apparently, ensue through two routes. One is through the release of $\beta\gamma$-subunits from activated G_i which then inhibit G_s dissociation and thus activation. The second appears to involve a direct inhibitory action of G_i on adenylate cyclase itself.

hydrolysable GTP analogues. These bind in a progressive fashion to the G-proteins and cause them to adopt a persistently activated state. This is due to the fact that normally the hydrolysis of GTP to GDP acts to 'turn-off' the functioning of these G-proteins.

'G_i' was originally defined as a species of G-protein which served to inhibit adenylate cyclase activity. This identification was made on the basis that pertussis toxin treatment of cells led to the abolition of receptor- and guanine-nucleotide-mediated inhibition of adenylate cyclase with the concomitant ADP-ribosylation of a 40 kDa G-protein α-subunit. It is now apparent, however, from molecular cloning, and biochemical and immunological studies, that there are at least three forms of G_i, called G_i1, G_i2 and G_i3 [5]. In addition, there is a highly related G-protein called G_z [6,7], which only differs from the 'G_i-family' in lacking the *C*-terminal cysteine residue that provides the ADP-ribosylation target, and also a further species called G_o, which can also be ADP-ribosylated by pertussis toxin [1,2]. While, neither G_z nor G_o are believed to be capable of inhibiting adenylate cyclase under normal conditions, it is still unclear as to whether all or just specific forms of G_i can fulfil such a role. Indeed, it has been suggested that one form of G_i may be the elusive pertussis toxin-sensitive 'G_p' [8], which can couple certain receptors to the stimulation of inositol phospholipid metabolism. Furthermore, some or all of the G_i species can couple receptors to K^+-channel opening and the activity of phospholipase A_2 may also be determined by the action of a member of the G_i family [9–11].

It is evident from this that changes in the concentration or activity of particular G-proteins may be expected to alter the responsiveness of adenylate cyclase to regulatory hormones and neurotransmitters. Indeed, patients who have Albright hereditary osteodystrophy exhibit a reduction in the mRNA coding for the α-subunit of G_s, show reduced G_s expression and resistance to hormones which act by stimulating adenylate cyclase activity [12,13].

One of our aims is to gain further understanding of the changes in cellular signal transduction systems that occur in diabetic and insulin-resistant states. To this end, we are particularly interested in the control of adenylate cyclase activity, where G-proteins play a pivotal role, and also in the functioning of the insulin receptor where we and, more recently, others have demonstrated an interaction with the G-protein systems which might account for some of the diverse actions of this hormone.

Diabetes-Induced Changes in Hepatocyte G-Protein Expression

Diabetes, induced by either streptozotocin or alloxan, causes a precipitate fall in plasma insulin concentrations and a concomitant dramatic elevation in blood glucose concentrations. It also leads to apparent insulin resistance in hepatocytes and an enhanced ability of glucagon to stimulate adenylate cyclase [14,15].

We have used anti-peptide antibodies and oligonucleotide probes, which are specific for particular G-proteins (Fig. 2), to determine both the expression of protein and mRNA of G_s and 'G_i' α-subunits and their associated β-subunits in hepatocytes of normal and streptozotocin-induced diabetic animals. Our studies show quite clearly that the induction of diabetes caused a reduction in the amount of both protein and mRNA for the α-subunits of G_i2 and G_i3, the two forms of 'G_i' expressed by hepatocytes (Fig. 3). Accompanying this is also a decrease in the amount of the $G_s\alpha$, as detected both by immunoblotting and Northern blot analysis.

These changes in expression were also seen using alloxan to induce diabetes and could be reversed by insulin therapy.

Such immunoblotting studies were performed on plasma membranes purified from isolated hepatocytes by a rapid and simple Percoll density gradient procedure [14]. Similar recoveries of plasma membranes were achieved using hepatocytes from either normal or diabetic animals and we observed no change in the buoyant density at which either adenylate cyclase or immunoreactive G_i co-migrated on the gradient.

Interestingly, Exton and co-workers [16] have reported their inability to confirm our original experiments showing that levels of G_i2 were reduced in diabetes. However, on close inspection of their methodology, it appears that they analysed a whole liver membrane preparation rather than one from hepatocytes. Now, hepatocytes contribute only some 60% of total liver cells, the rest being Kupffer cells, macrophages and various other blood-derived cells [17,18]. So, we have performed parallel experiments where we obtained mRNA from whole liver and also used the 'Neville-type' of method to prepare plasma membrane fractions from whole liver as described by Exton and co-workers. In these studies we were able to reproduce the results of Exton and co-workers [16], who found that diabetes elicited no change in G_i2 levels and

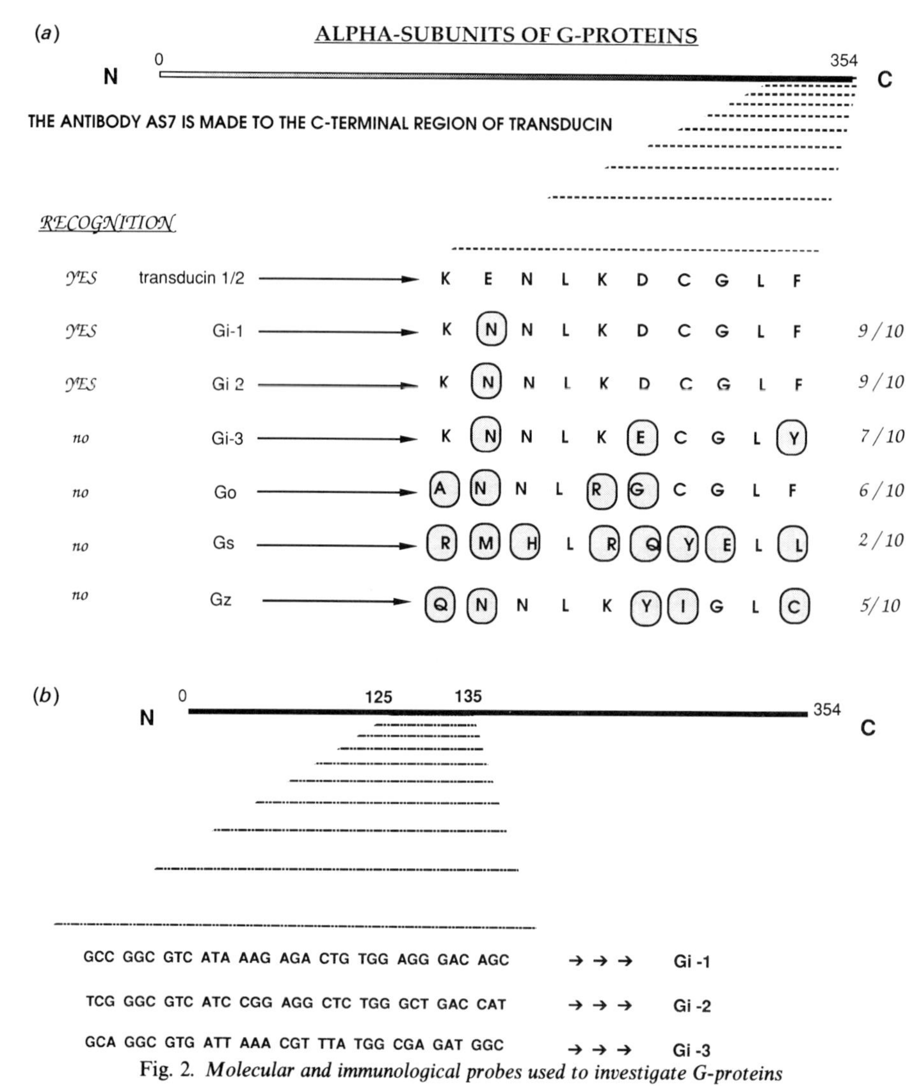

Fig. 2. *Molecular and immunological probes used to investigate G-proteins*

(*a*) The specificity of one of the anti-peptide antibodies (AS7) that we have used is shown. This was raised against the *C*-terminal decapeptide of the G-protein transducin, which is only found in retinal rods and cones. The sites and of differences for residues of related G-proteins are shown as are the 'scores' for similarity. (*b*) The sequence of the synthetic oligonucleotide probes which we have found to identify specific G-protein messages is shown.

actually a small increase in the levels of G_s (Fig. 3). This suggests that diabetes exerts very different effects on G-protein expression in the non-parenchymal cells of the liver compared with the parenchymal cells (hepatocytes). Analysis of transcripts in whole liver showed that the induction of diabetes exerted very little change in those for $G_i\alpha$ forms and a small increase in that for $G_s\alpha$. Thus, the diabetes-induced changes in hepatocyte G-protein transcripts are presumably diluted out by transcripts accruing from the non-parenchymal cells when analysis is performed upon whole liver mRNA. It seems likely that this also happens for immunoblotting analyses done on whole liver plasma membrane

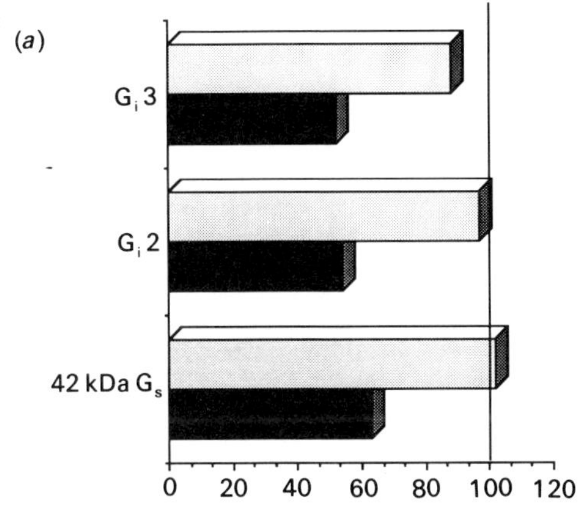

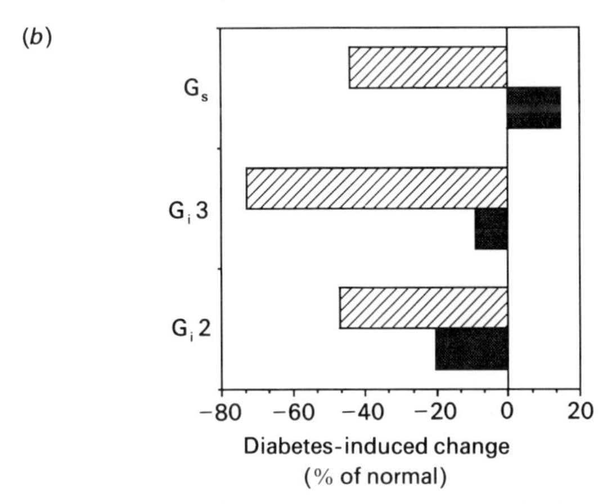

Fig. 3. *Diabetes causes different changes in G-protein α-subunit protein and transcripts in whole liver compared to hepatocytes*

This compares analyses of whole liver (▨) and isolated hepatocytes (■) in immunoblotting studies for specific G-protein α-subunits (*a*) and for transcripts using synthetic oligonucleotide probes in Northern blot analyses (*b*); ▨, hepatocytes; ■, whole liver.

preparations. Indeed, this may be further complicated by the possibility that the membrane preparation analysed by Exton and co-workers [16] may actually be enriched in plasma membranes accruing from non-parenchymal cells.

Diabetes-Induced Changes in Adipocyte G-Protein Expression

In contrast to the situation with hepatocytes, streptozotocin-induced diabetes did not decrease the amount of G-protein α-subunits in adipocyte plasma membranes [19]. Indeed, while the amounts of $G_i\alpha$, $G_i2\alpha$ and $G_s\alpha$ were unchanged, the amount of $G_i3\alpha$ was nearly doubled in membranes from diabetic animals (Fig. 4*a*). Analysis of transcripts showed that diabetes caused no alterations in those for the α-subunits of either G_i2 or G_s, but dramatically increased the number encoding both $G_i1\alpha$ and $G_i3\alpha$ (Fig. 4*b*).

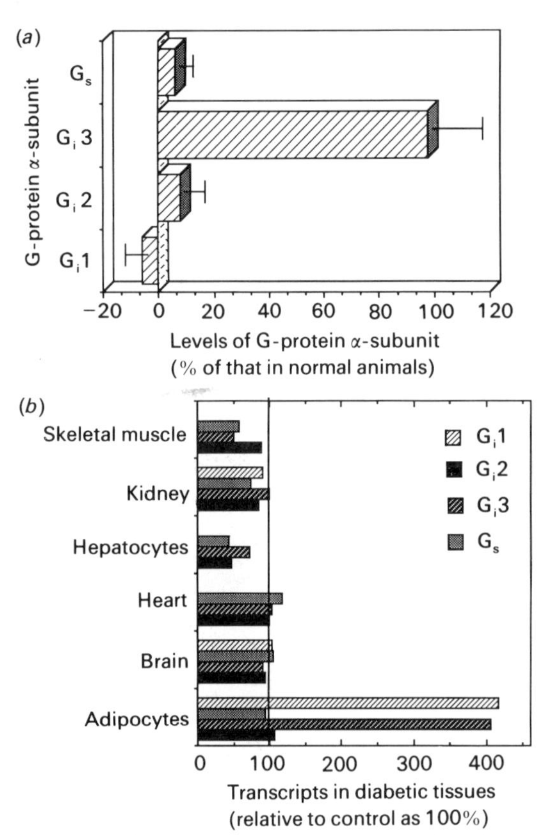

Fig. 4. *Diabetes-induced changes in G-protein α-subunit protein and transcripts in adipocytes*
Immunoblotting studies for specific G-protein α-subunits (*a*) are shown for adipocyte membranes and in (*b*) transcript analyses for both adipocytes and other tissues/cells.

That such a marked increase in the transcripts for the α-subunits of G_i1 and $G_i3\alpha$ was not accompanied by any proportional increase in membrane-associated protein may be due to the fact that diabetes had little effect upon the number of β-subunit transcripts. Thus, it could be that for changes in the number of plasma membrane G-protein α-subunits to occur, they need to be stabilized by association with their corresponding β-subunits. Certainly, such an interpretation is consistent with recent studies which have shown that the persistent activation, and thus dissociation, of G_s in intact cells leads to the down-regulation (loss) of $G_s\alpha$ from the plasma membrane.

Diabetes-Induced Changes in G-Protein Transcripts in other Tissues

We have also determined G-protein transcripts in brain, kidney, heart and skeletal muscle (Fig. 4*b*). In contrast to the changes seen in hepatocytes and adipocytes, diabetes elicited little alteration in G-protein transcripts in brain, kidney, skeletal muscle and heart. However, all such tissues contain multiple cell types and it is possible that this type of analysis could mask more subtle changes going on in specific cell types.

It is clear, nevertheless, from our studies that diabetes causes tissue-selective changes in the expression of particular G-protein α-subunits. Such actions appear to be most pronounced in two cell types which provide key targets for the action of insulin, namely hepatocytes and adipocytes.

Diabetes-Induced Changes in G_i-Mediated Inhibition of Adenylate Cyclase

G_i-mediated inhibition of adenylate cyclase can be assessed using two distinct strategies (Fig. 1). One involves the occupancy and subsequent action of an appropriate inhibitory receptor. The other involves the direct activation of G_i itself using guanine nucleotides to effect inhibition and thus to 'short-circuit' or by-pass the action of the receptor. In the latter instance, this is often effected by using non-hydrolysable guanine nucleotides which can, at low concentrations, selectively activate G_i because this G-protein exhibits a higher affinity for them than does G_s. However, GTP itself can also be used. In this case, the degree of inhibition seen is less than that achieved when coupled to a suitable inhibitory receptor and it also occurs at GTP concentrations which are much higher than those required to satisfy the receptor-mediated stimulation of G_s. Nevertheless, it should be appreciated that this GTP-mediated inhibition is likely to be seen under physiological conditions. This is due to the fact that intracellular GTP concentrations in at least hepatocytes and adipocytes are of the order of 600 μM and can thus be expected to be responsible for causing a G_i-mediated tonic inhibitory effect upon adenylate cyclase activity under normal and stimulated conditions.

The induction of diabetes quite clearly (Fig. 5) abolishes the guanine-nucleotide-mediated inhibition of adenylate cyclase in plasma membranes from both adipocytes and hepatocytes [15,19]. This removes the tonic inhibitory action of 'G_i' on adenylate cyclase and causes an increase not only in its

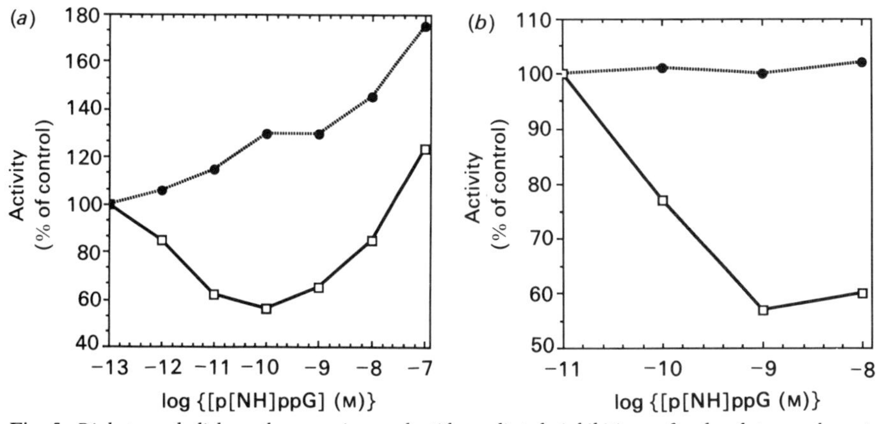

Fig. 5. *Diabetes abolishes the guanine-nucleotide-mediated inhibition of adenylate cyclase in adipocytes (a) and hepatocytes (b)*

'G_i' function is here assessed by first amplifying the basal activity of adenylate cyclase with the diterpene forskolin and then using low concentrations of the GTP analogue guanosine 5′-[β,γ]-imidodiphosphate (p[NH]ppG) to activate G_i. Higher concentrations lead to the activation of G_s. ●, Diabetic, □; normal.

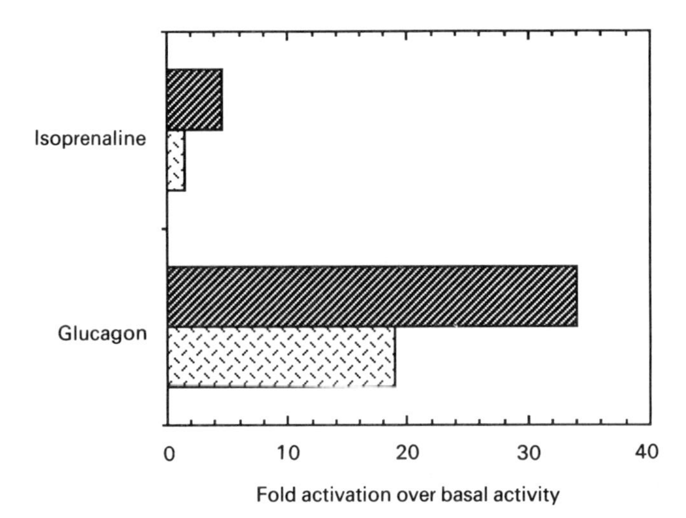

Fig. 6. *Enhanced action of stimulatory hormones on adenylate cyclase activity in adipocytes and hepatocytes*

This is demonstrated for isoprenaline in adipocytes and glucagon in hepatocytes at maximally effective concentrations of these ligands. ▨, Diabetic; ⊡, normal.

GTP-dependent activity but most evidently in its hormone-stimulated activity, achieved by glucagon in hepatocytes and β-adrenoceptor activation in adipocytes (Fig. 6).

That such a potentiation of hormonal stimulation of adenylate cyclase is achieved through the loss of the tonic inhibitory function of G_i receives

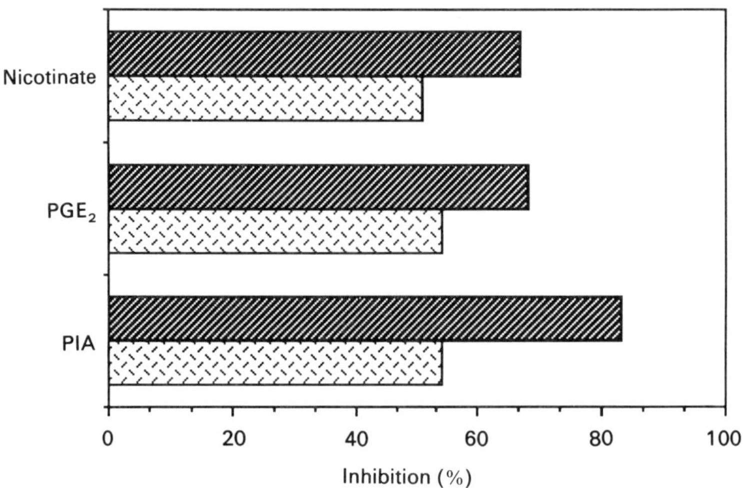

Fig. 7. *Continued action of receptor-mediated inhibition in adipocytes and hepatocytes of diabetic animals*

Demonstrates that receptor inhibition of adipocyte adenylate cyclase is not abolished by the induction of diabetes. That inhibition in the diabetic state seems larger than in the control is more apparent than real. This is because such studies have to be done in the presence of GTP; thus, in cells from normal animals a degree of GTP-mediated inhibition has already occurred in the absence of receptor ligand. This is lost in the diabetic state. ▨, Diabetic; ⊡, normal.

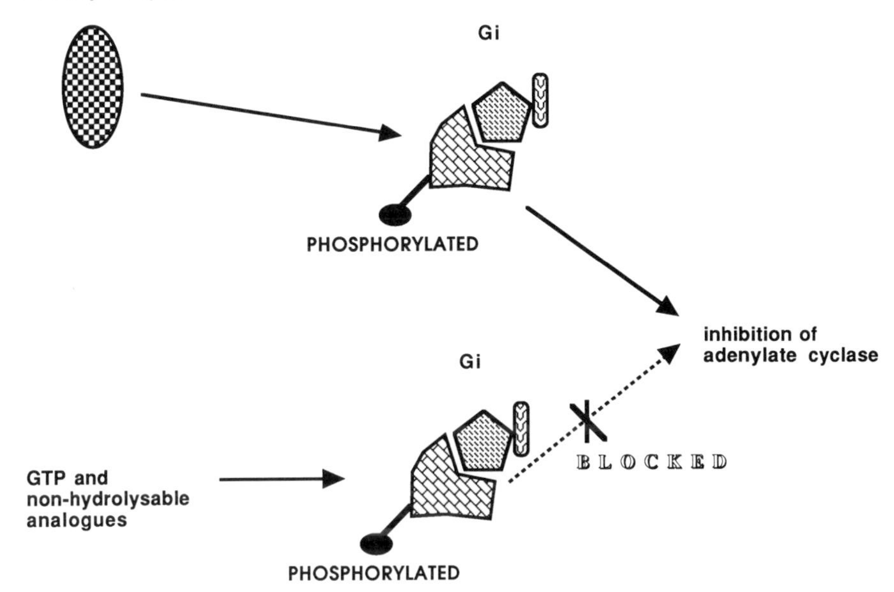

Fig. 8. *Scheme for the selective action of diabetes on G_i functioning in adipocytes and hepatocytes* This suggests that the phosphorylation of $G_i2\alpha$ leads to a conformational change which stops the GTP-mediated inhibitory action of G_i. We presume that the physical interaction of the receptor with G_i exerts a sufficiently powerful influence to exert its own conformational change which overcomes that induced by phosphorylation.

support from our experiments using pertussis toxin to treat cells from normal animals. This leads to the inactivation of G_i and similar potentiation in functioning of stimulatory receptors.

The pathophysiological consequence of this in diabetes is the loss of the tonic inhibitory function of G_i. This may be expected to exacerbate the diabetic state by potentiating the gluconeogenic and ketogenic action of glucagon and the β-adrenoceptor-mediated action of adrenaline.

Interestingly, however, while diabetes causes the loss of the tonic inhibitory function of G_i, it does not abolish the receptor-mediated functioning of G_i [19]. Thus, in adipocytes, the functioning of inhibitory nicotinate and adenosine receptors persists (Fig. 7), as does the functioning of the P_{2y}-purinoceptor in hepatocytes from diabetic animals. In the latter instance, the effectiveness was reduced by some 50%, presumably due to the fall in expression of $G_i2\alpha$. The induction of diabetes can thus be seen to cause a selective crippling of the functioning of this inhibitory G-protein.

The Molecular Basis of Diabetes-Induced Changes in G_i Functioning

Changes in the expression of particular forms of G_i cannot provide the basis for the total loss of guanine-nucleotide-mediated 'G_i function' seen in both the adipocytes and hepatocytes of diabetic animals [15,19]. In both cell types guanine nucleotide-mediated inhibitory action is lost, yet it is only in hepatocytes where G_i expression is reduced and even then it is not abolished. Furthermore, while diabetes causes loss of the guanine-nucleotide-mediated inhibitory action of G_i, it does not abolish the receptor-mediated inhibitory action of G_i.

We would suggest that it is rather unlikely that two distinct G-proteins mediate each of these individual effects, i.e. one only activated by guanine nucleotides and the other insensitive to either non-hydrolysable guanine nucleotides or GTP and only activated by receptor interaction. Rather, such observations might be rationalized by assuming that one (or more) form of G_i became reversibly modified in the diabetic animals, such that it became insensitive to the sole application of guanine nucleotides (perhaps a rise in K_m), yet the conformational change elicited by such a modification could be overcome when the G-protein interacts with an appropriate receptor (Fig. 8).

We believe, as detailed below, that the molecular basis for this event is the phosphorylation of the α-subunit of G_i2. Indeed, we have shown that such a mechanism is available in intact hepatocytes [20,21].

The Selective Phosphorylation of G_i2

The phosphorylation of the α-subunit of a solubilized, purified G_i preparation by purified protein kinase C (C-kinase) was first demonstrated by Jakobs and co-workers [22]. More recently, both the purified human insulin receptor tyrosyl kinase [23,24] and cyclic AMP-dependent protein kinase (A-kinase) [25] have been shown to phosphorylate the α-subunit of purified, solubilized G_i preparations. The phosphorylation of G_i was suggested, by Jakobs and co-workers [22], to lead to inactivation of G_i because α_2-adrenergic inhibition of platelet adenylate cyclase was blocked by treatment with 12-*O*-tetradecanoyl phorbol 13-acetate (TPA), a phorbol ester which activates protein kinase C. However, C-kinase has been shown to cause the phosphorylation and inactivation of a number of receptors, including various adrenoceptors [26]. It is thus unclear as to whether TPA was exerting its inhibitory effect on α_2-adrenoceptor function by either receptor phosphorylation or a presumed action on 'G_i'.

We have demonstrated that challenge of intact hepatocytes with either glucagon, angiotensin or vasopressin, but not insulin, leads to the specific phosphorylation of $G_i2\alpha$ and not $G_i3\alpha$ or $G_s\alpha$ [20,21] (Fig. 9). This can readily be seen by the specific immunoprecipitation of these G-protein α-subunits from intact hepatocytes which had been preincubated with $[^{32}P]P_i$ before hormonal challenge (Fig. 9). Thus the anti-peptide antisera AS7 and I3B were shown to immunoprecipitate G_i2 and G_i3 selectively and the antiserum CS1 immunoprecipitated the stimulatory G-protein, G_s. Hepatocytes do not appear to express G_i1. This is based both on our failure to detect immunoreactive material in hepatocyte plasma membranes using a specific anti-peptide antiserum which detects G_i1 in both adipocytes and brain and the detection of marginal levels of transcripts for this G-protein in hepatocytes. It would thus

(a)

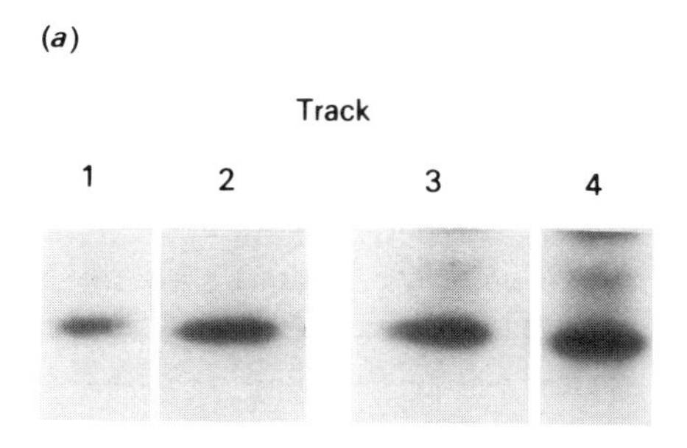

(b)

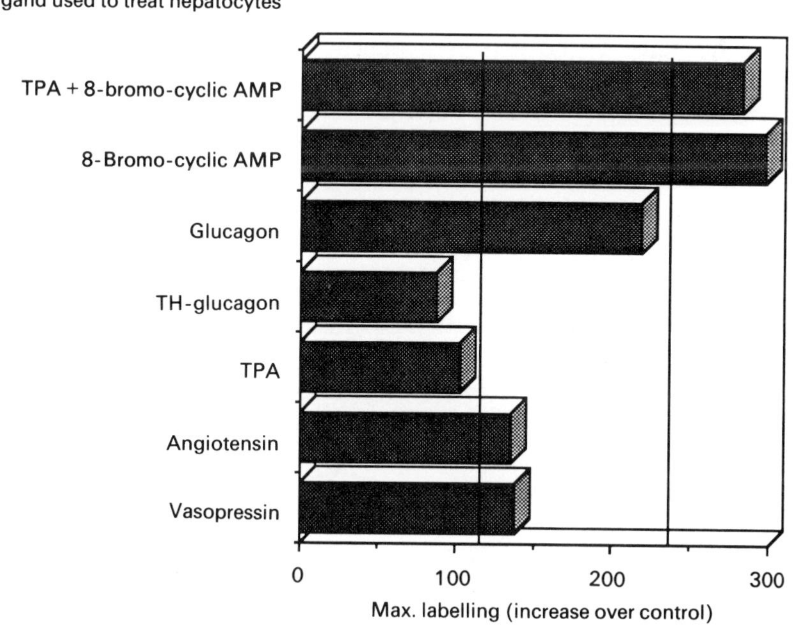

Fig. 9. *Phosphorylation of $G_i2\alpha$ in intact hepatocytes*

(a) The immunoprecipitation of labelled $G_i2\alpha$ under basal conditions (tracks 1, 3) and when stimulated by the phorbol ester TPA (tracks 2, 4) is shown. Tracks 1 and 2 are data from normal hepatocytes and tracks 3 and 4 from the hepatocytes of diabetic animals. Note that TPA only enhances phosphorylation in hepatocytes from normal animals. In (b) the relative increases in labelling of $G_i2\alpha$ achieved by various agents are shown. Note that glucagon and 8-bromo-cyclic AMP appear to catalyse the incorporation of twice as much phosphate as the others. We suggest that there are two sites for phosphorylation on $G_i2\alpha$, one being a target for C-kinase and the other mediated either directly or indirectly by A-kinase.

seem then that the inhibition of adenylate cyclase in hepatocytes must be mediated through the action of either G_i2 or G_i3.

The immunoprecipitation of phosphorylated $G_i2\alpha$, by the antiserum AS7, was blocked in a dose-dependent fashion by the inclusion of the *C*-terminal decapeptide of transducin in the immunoprecipitation assay. This, together with our failure to immunoprecipitate this species with pre-immune serum, demonstrated that the phosphorylated species was not non-specifically immunoprecipitated. This was further supported by the fact that inclusion, in the immunoprecipitation assay, of the *C*-terminal decapeptide of G_z [6,7], a G-protein which is related to G_i, but has no site for pertussis toxin-catalysed ADP-ribosylation, failed to block the immunoprecipitation of the labelled species (Fig. 10). This, we consider, is an important experiment as G_z is believed to be phosphorylated in human platelets [27] and we felt it important to demonstrate that the labelled species which we have detected was not the α-subunit of G_z.

The antiserum AS7 will only detect $G_i1\alpha$ and $G_i2\alpha$. However, as hepatocytes express $G_i2\alpha$, but not $G_i1\alpha$, we firmly believe that the labelled species immunoprecipitated is $G_i2\alpha$. Indeed, we have confirmed such data with an antibody which is specific for $G_i2\alpha$ and does not detect $G_i1\alpha$.

Over the same period that pretreatment with the various ligands caused ^{32}P to be incorporated into $G_i2\alpha$, we observed a parallel loss in the ability of guanine nucleotides to inhibit adenylate cyclase activity in isolated plasma membranes, i.e. 'G_i function' (Fig. 11). In this regard, the state engendered by ligand-mediated phosphorylation of $G_i2\alpha$ paralleled that seen in membranes from diabetic animals [15,19].

Such phosphorylation and inactivation of G_i was also achieved by treating the hepatocytes with the tumour-promoting phorbol ester TPA [20,21]. This

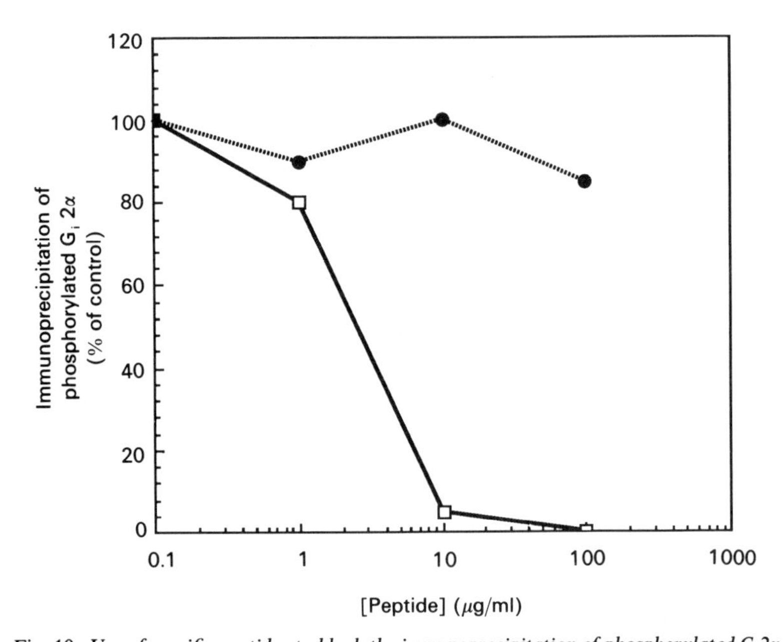

Fig. 10. *Use of specific peptides to block the immunoprecipitation of phosphorylated $G_i2\alpha$*

To the immunoprecipitation assays were added either the *C*-terminal decapeptide of transducin (□) or of G_z (●). We see that only in the case of the transducin peptide was immunoprecipitation blocked. Thus $G_i2\alpha$ is specifically immunoprecipitated as a phosphoprotein. Sequences are given in Fig. 2.

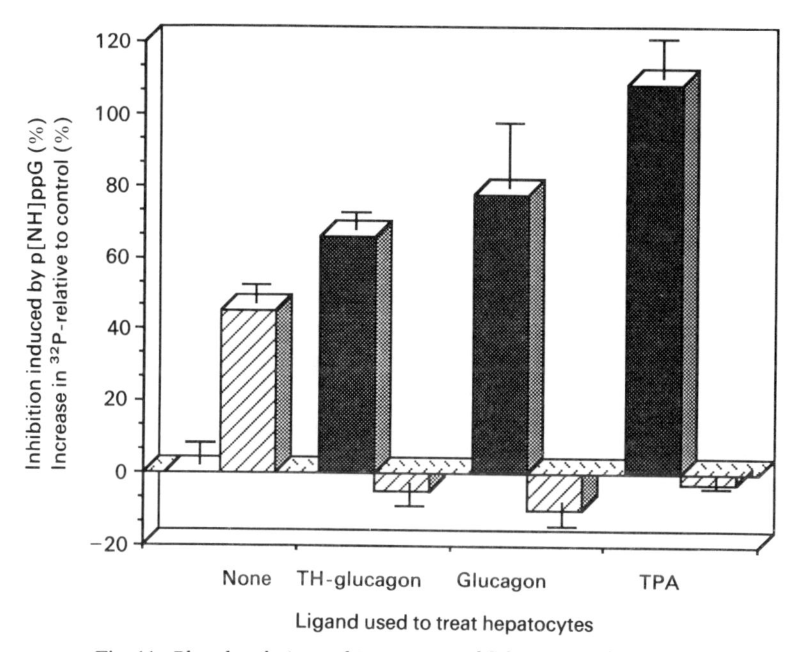

Fig. 11. *Phosphorylation and inactivation of $G_i2\alpha$ in intact hepatocytes*

This demonstrates how changes in phosphorylation of $G_i2\alpha$ are linked to the loss of 'G_i' function. This can be further confirmed by dose–effect and time-course studies. ■, Phosphorylation; ▨, inhibition.

agent mimics the structure of diacylglycerol and is capable of exerting a potent stimulatory effect upon protein kinase C in a variety of cells, including hepatocytes. Indeed, both angiotensin and vasopressin also activate protein kinase C in hepatocytes as they serve to stimulate inositol phospholipid metabolism and thus cause the production of diacylglycerol [28–30]. In this regard, although glucagon activates adenylate cyclase in hepatocytes, there is also substantial evidence to show that it can elevate intracellular Ca^{2+} and can exert cyclic AMP-independent actions involving the activation of protein kinase C [31–39]. This it achieves by elevating intracellular diacylglycerol concentrations, in part by eliciting a small stimulation of inositol phospholipid metabolism and also by stimulating the breakdown of phosphatidylcholine. Certainly this is consistent with our observation that the glucagon analogue, TH-glucagon, can stimulate the phosphorylation of $G_i2\alpha$ and increase intracellular diacylglycerol concentrations, but does not increase intracellular cyclic AMP concentrations.

Thus, challenge of intact hepatocytes with vasopressin, TH-glucagon or TPA led to a rapid, time-dependent increase in the labelling of $G_i2\alpha$ which paralleled the loss of 'G_i function'. Maximal effects occurred within 5 min of addition of each ligand. Similarly, treatment of hepatocytes with glucagon led to the rapid loss of 'G_i function', although maximal labelling was achieved after slightly longer periods of time. Hepatocytes challenged with 8-bromo-cyclic AMP also displayed increases in both labelling and loss of 'G_i

function'; however, these events occurred at a much slower rate, with maximal effects being achieved after some 15 min. Dose–effect curves for both the ligand-stimulated increases in the labelling of $G_i2\alpha$ and the loss of 'G_i function' were similar when cells were challenged with TPA [EC_{50} (concentration at which 50% of maximal effect is achieved) = 0.4–0.7 nM], vasopressin (EC_{50} = 0.7–1.1 nM), angiotensin II (EC_{50} = 0.2–0.5 nM), TH-glucagon (EC_{50} = 0.36–0.75 nM) and 8-bromo cyclic AMP (EC_{50} = 1.8–4 μM). In contrast to the monophasic dose–effect curves seen with these agents, the dose–effect curve for the glucagon-mediated increase in the labelling of $G_i2\alpha$ was markedly biphasic with EC_{50} values for the two components of around 36 pM and 20 nM. The loss of 'G_i function' paralleled the high-affinity component of glucagon's labelling of $G_i2\alpha$ (EC_{50} = 25 pM). TPA, TH-glucagon, angiotensin II and vasopressin achieved similar maximal increases in the labelling of $G_i2\alpha$; these ranged from 90 to 140% over that observed in the basal state. In contrast, treatment of hepatocytes with either glucagon (1 μM) or 8-bromocyclic AMP (100 μM) caused a maximal increase in labelling of around 220–300%, while the increase in labelling achieved by nanomolar glucagon concentrations was around 100%. Analysis of the phosphoamino acid content of immunoprecipitated $G_i2\alpha$ from cells treated with either TPA or 8-bromocyclic AMP showed phosphoserine only. Incubation of hepatocyte membranes with [γ-^{32}P]ATP and purified protein kinase C, but not protein kinase A, led to the incorporation of label into immunoprecipitated $G_i2\alpha$. This labelling was abolished if membranes were obtained from cells which had received prior treatment with ligands shown to cause the phosphorylation of $G_i2\alpha$ in intact cells.

We thus believe that in intact hepatocytes $G_i2\alpha$ can be phosphorylated by protein kinase C and that this causes the loss of guanine-nucleotide-mediated inhibition of adenylate cyclase. This indicates that it is G_i2 which mediates the guanine-nucleotide-mediated inhibition of adenylate cyclase in intact hepatocytes.

In support of our intact cell studies, we found [21] that treatment of isolated hepatocyte membranes with pure protein kinase C and [γ^{32}P]ATP also led to the phosphorylation of $G_i2\alpha$. Intriguingly, however, if hepatocyte membranes from diabetic animals were used, then no phosphorylation could be achieved by subsequent treatment with pure protein kinase C unless these membranes had been pretreated with alkaline phosphatase. This suggested to us that the crippling of G_i function in diabetes is due to the phosphorylation of $G_i2\alpha$. This may be because in streptozotocin-induced diabetes, plasma vasopressin levels become markedly elevated and can thus be expected to stimulate hepatocyte inositol phospholipid metabolism, leading to the activation of protein kinase C and the concomitant phosphorylation of $G_i2\alpha$.

We have noted that some labelling of $G_i2\alpha$ occurred under basal conditions in hepatocytes from normal animals. However, employing hepatocytes from diabetic animals we found that, upon incubation with [^{32}P]P_i, then considerable labelling was evident. Indeed, from estimates of the amount of $G_i2\alpha$ present together with the specific radioactivity of the intracellular ATP pool, we assess that $G_i\alpha$ is labelled to an approximate stoichiometry of approximately 1.2 mol of P_i/mol of protein in hepatocytes from diabetic animals. This compares with around 0.3 mol of P_i/mol of $G_i2\alpha$ under basal conditions for hepatocytes of normal animals.

The Functioning and Phosphorylation of G_i in Hepatocytes of Zucker Rats

Zucker rats can be identified as normal, lean animals or as obese littermates [40]. The obese animals at some 8–10 weeks of age exhibit dramatically elevated blood insulin levels compared with their lean littermates, but exhibit normal blood glucose levels. The obese animals thus provide a model of insulin resistance as is seen associated with obesity rather than diabetes *per se* [14,40].

Nevertheless, we have found that hepatocyte membranes only from the lean and not the obese animals exhibited guanine-nucleotide-controlled G_i function [41]. This is not accompanied by any changes in the levels of expression of either $G_i2\alpha$ or $G_i3\alpha$, although the levels of $G_s\alpha$ fall by some 50% in membranes from the obese animals compared with their lean littermates.

We believe that the molecular basis for the crippled G_i functioning in hepatocytes from the obese animals is exactly the same as is seen in streptozotocin-induced diabetic rats. Thus, challenge of hepatocytes with glucagon, TH-glucagon, TPA or angiotensin causes the phosphorylation of $G_i2\alpha$ in cells from lean, but not obese, animals; treatment of hepatocyte membranes from lean, but not obese, animals with protein kinase C causes the phosphorylation of $G_i2\alpha$ unless the membranes from the obese animals had been pretreated with alkaline phosphatase and, finally, incubation of hepatocytes from obese animals with $[^{32}P]P_i$ led to the slow incorporation of label up to a stoichiometry of approximately 1 mol of P_i/mol of $G_i2\alpha$ compared with about 0.5 mol of P_i/mol of $G_i2\alpha$ in their lean counterparts.

Conclusions

Our studies show that (Fig. 12): (i) the induction of diabetes leads to tissue-specific changes in the expression of selected G-proteins; (ii) a major feature which appears to characterize diabetic and associated insulin-resistant states is a crippled functioning of G_i: this manifests itself as a loss of the tonic, GTP-mediated inhibitory action of this G-protein, leading to a small change in the basal activity of adenylate cyclase, but, most evidently, to a large potentiation of the action of stimulatory hormones on this enzyme; this amplification of the action of stimulatory hormones is consistent with similar actions mediated by the inactivation of G_i using pertussis toxin treatment; (iii) G_i2 appears to mediate the inhibitory effect of guanine nucleotide on adenylate cyclase; and (iv) $G_i2\alpha$ can be phosphorylated by the action of protein kinase C and this leads to a loss in its ability to mediate the tonic guanine-nucleotide-elicited inhibition of adenylate cyclase.

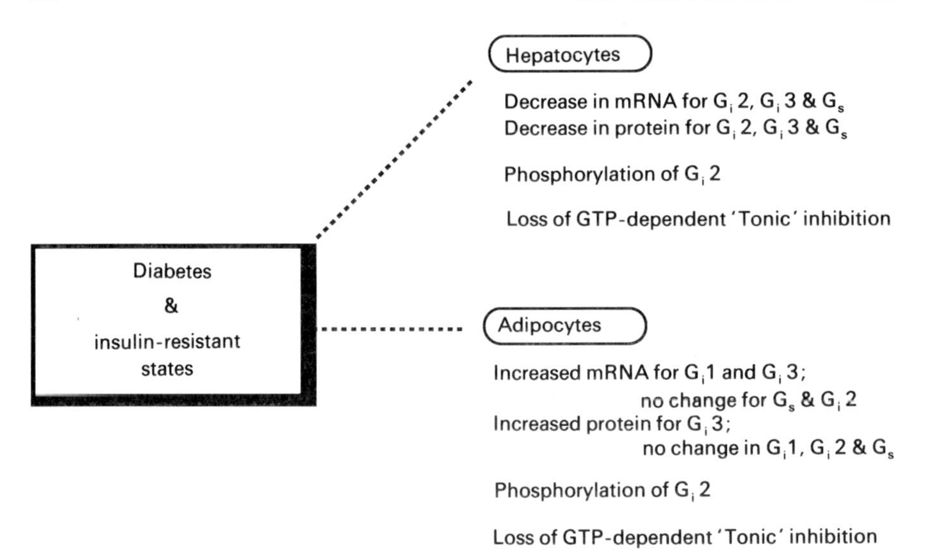

Fig. 12. *A schematic summary of changes in adipocyte and hepatocyte G-protein expression and functioning in insulin-resistant states*

That hepatocytes from obese but not lean Zucker rats appear to show the same defect as do adipocytes from diabetic but not normal *db/db* mice, suggests that this altered functioning of $G_i2\alpha$ might characterize insulin-resistant states. This clearly has profound consequences for the functioning of adenylate cyclase and receptors that serve to stimulate it. However, there is also evidence that pertussis toxin treatment can block certain of insulin's actions [42–44], which suggest that the insulin receptor may interact with a pertussis toxin-sensitive G-protein. That the insulin receptor may interact with G_i has gained support from observations that insulin can attenuate the ability of pertussis toxin to cause the ADP-ribosylation of G_i2 in hepatocytes [45,46] and that at least the purified insulin receptor can cause the phosphorylation of both purified G_i and G_o while they are in their GDP-bound state [23,24].

It is clear that the expression and functioning of members of the G-protein family can be severely disturbed in both diabetic and insulin-resistant states. In the case of G_i, it remains to be established whether the altered function ensues as a result of the action of the same process which leads to insulin resistance, or whether such a modification of G_i actually contributes itself to forming the insulin-resistant state. Interestingly, there has long been thought to be a link between insulin resistance and hypertension and recently it has been shown that G_i function is reduced in the platelets of spontaneously hypertensive rats [47]. (Indeed, one group of investigators have shown that G_i can be phosphorylated in human platelets through C-kinase action [48].)

We thank the M.R.C., A.F.R.C., B.D.A. and California Metabolic Research Foundation for generous financial support.

References

1. Gilman, A. G. (1987) *Annu. Rev. Biochem.* **56**, 615–649
2. Milligan, G. (1988) *Biochem. J.* **255**, 1–13
3. Birnbaumer, L., Codina, J., Mattera, R., Cerione, R. A., Hildebrandt, J. D., Sunyer, T., Caron, M. G., Lefkowitz, R. J. & Iyengar, R. (1985) *Mol. Aspects Cell. Regul.* **4**, 131–182
4. Houslay, M. D. (1984) *Trends Biochem. Sci.* **9**, 615–649
5. Jones, D. T. & Reed, R. R. (1987) *J. Biol. Chem.* **262**, 14241–14249
6. Fong, H. K. W., Yoshimoto, K. K., Eversole-Cire, P. & Simon, M. (1988) *Proc. Natl. Acad. Sci. U.S.A.* **85**, 3066–3070
7. Matsoka, M., Itoh, H., Kozasa, T. & Kaziro, Y. (1988) *Proc. Natl. Acad. Sci. U.S.A.* **85**, 5384–5388
8. Cockroft, S. (1987) *Trends Biochem. Sci.* **12**, 75–78
9. Jelsema, C. L. & Axelrod, J. (1987) *Proc. Natl. Acad. Sci. U.S.A.* **84**, 3623–3627
10. Yatani, A., Codina, J., Brown, A. M. & Birnbaumer, L. (1987) *Science* **235**, 207–211
11. Yatani, A., Mattera, R., Codina, J., Gref, R., Okabe, K., Padrell, E., Iyengar, R., Brown, A. M. & Birnbaumer, L. (1988) *Nature* (*London*) **336**, 680–682
12. Motulsky, H. J., Hughes, R. J., Brickman, A. S., Farfel, K. D., Bourrne, H. R. & Insel, P. A. (1982) *Proc. Natl. Acad. Sci. U.S.A.* **79**, 4193–4197
13. Levine, M. A., Ahn, T. G., Klupt, S. F., Kaufman, K. D., Smallwood, P. M., Bourne, H. R., Sullivan, K. A. & Van Dop, C. (1988) *Proc. Natl. Acad. Sci. U.S.A.* **85**, 617–621
14. Gawler, D., Wilson, A. & Houslay, M. D. (1989) *J. Endocrinol.* **122**, 207–212
15. Gawler, D., Milligan, G., Spiegel, A. M., Unson, C. G. & Houslay, M. D. (1987) *Nature* (*London*) **327**, 229–232
16. Lynch, C. J., Blackmore, P. F., Johnson, E. H., Wange, R. L., Krone, P. K. & Exton, J. H. (1989) *J. Clin. Invest.* **83**, 2050–2062
17. Jungermann, K. & Sass, D. (1987) *Trends Biochem. Sci.* **12**, 198–202
18. Van Berkel, T. J. C. (1979) *Trends Biochem. Sci.* **5**, 202–205
19. Strassheim, D., Milligan, G. & Houslay, M. D. (1990) *Biochem. J.* **266**, 521–526
20. Pyne, N. J., Murphy, G. J., Milligan, G. & Houslay, M. D. (1989) *FEBS Lett.* **243**, 77–82
21. Bushfield, M., Murphy, G. J., Parker, P. J., Hruby, V. J., Milligan, G. & Houslay, M. D. (1990) *Biochem J.* in the press
22. Katada, T., Gilman, A. G., Watanabe, Y., Bauer, S. & Jakobs, K. H. (1985) *Eur. J. Biochem.* **151**, 431–437
23. O'Brien, R. M., Houslay, M. D., Milligan, G. & Siddle, K. (1987) *FEBS Lett.* **212**, 281–288
24. Krupinski, J., Rajaram, R., Lakonuhok, M., Benovic, J. L. & Cerione, R. A. (1988) *J. Biol. Chem.* **263**, 12333–12341
25. Watanabe, Y., Imaizumi, T., Misaki, N., Iwakura, K. & Yoshida, H. (1988) *FEBS Lett.* **236**, 372–374
26. Houslay, M. D. (1989) *Curr. Opin. Cell Biol.* **1**, 669–674
27. Carlson, K. E., Brass, L. F. & Manning, D. R. (1989) *J. Biol. Chem.* **264**, 13298–13305
28. Creba, J. A., Downes, C. P., Hawkins, P. T., Brewster, G., Michell, R. H. & Kirk, C. J. (1983) *Biochem. J.* **212**, 733–747
29. Bocckino, S. B., Blackmore, P. F. & Exton, J. H. (1985) *J. Biol. Chem.* **260**, 14201–14207
30. Garrison, J. C., Johnsen, D. E. & Campanile, C. P. (1984) *J. Biol. Chem.* **259**, 3283–3294
31. Heyworth, C. M. & Houslay, M. D. (1983) *Biochem. J.* **214**, 93–98
32. Heyworth, C. M., Wallace, A. V. & Houslay, M. D. (1983) *Biochem. J.* **214**, 99–110
33. Wakelam, M. J. O., Murphy, G. J., Hruby, V. J. & Houslay, M. D. (1986) *Nature* (*London*) **323**, 68–71
34. Murphy, G. J., Hruby, V. J., Trivedi, D., Wakelam, M. J. O. & Houslay, M. D. (1987) *Biochem. J.* **243**, 39–46
35. Heyworth, C. M., Whetton, A. D., Kinsella, A. R. & Houslay, M. D. (1984) *FEBS Lett.* **170**, 38–42
36. Murphy, G. J., Gawler, D., Milligan, G., Wakelam, M. J. O., Pyne, N. J. & Houslay, M. D. (1989) *Biochem. J.* **259**, 191–197
37. Whipps, D. E., Armstron, A. E., Pryor, H. J. & Halestrap, A. P. (1987) *Biochem. J.* **241**, 835–845
38. Blackmore, P. F. & Exton, J. H. (1986) *J. Biol. Chem.* **261**, 11056–11063
39. Bocckino, S. B., Blackmore, P. F. & Exton, J. H. (1985) *J. Biol. Chem.* **260**, 14201–14207
40. Bray, G. A. (1977) *Fed. Proc. Fed. Am. Soc. Exp. Biol.* **36**, 148–153
41. Houslay, M. D., Gawler, D. J., Milligan, G. & Wilson, A. (1989) *Cellular Signalling* **1**, 9–22
42. Houslay, M. D. (1986) *Biochem. Soc. Trans.* **14**, 183–193

43. Heyworth, C. M., Hanski, E. & Houslay, M. D. (1983) *Biochem. J.* **222**, 189–194
44. Elks, M. L., Watkins, P. A., Manganiello, V. C., Moss, J., Hewlett, E. & Vaughan, M. (1983) *Biochem. Biophys. Res. Commun.* **116**, 593–596
45. Pyne, N. J., Heyworth, C. M., Balfour, N. W. & Houslay, M. D. (1990) *Biochem. Biophys. Res. Commun.* **165**, 251–256
46. Rothenberg, P. L. & Kahn, C. R. (1988) *J. Biol. Chem.* **263**, 15546–15552
47. Coquil, J. F. & Brunelle, G. (1989) *Biochem. Biophys. Res. Commun.* **162**, 1256–1271
48. Crouch, M. F. & Lapetina, E. G. (1988) *J. Biol. Chem.* **263**, 3363–3371

Biochem. Soc. Symp. **56**, 155–164
Printed in Great Britain

Regulation of Transmembrane Signalling Elements: Transcriptional, Post-Transcriptional and Post-Translational Controls

CRAIG C. MALBON, JOHN R. HADCOCK, PETER J. RAPIEJKO, MANUEL ROS, HSIEN-YU WANG and DAVID C. WATKINS

Department of Pharmacology, Diabetes and Metabolic Diseases Research Program, State University of New York at Stony Brook, NY 11794, U.S.A.

Synopsis

G-protein-mediated transmembrane signalling is a common motif in biology. The actions of a populous group of G-protein-linked receptors in hormone action, olfaction and vision in vertebrates are examples in which input signals are transferred from a receptor molecule (or photopigment) to an effector unit(s) via G-proteins. The expression and functional status of the receptors, G-proteins, and effectors that constitute these transmembrane signalling systems are regulated physiologically. Altering the abundance, function, or both of these elements provides the means for modulating transmembrane signalling and integration of information among separate pathways. Recent advances in the cell and molecular biology of transmembrane signalling elements provide insight as to the mechanisms by which regulation occurs. Transcriptional control is exemplified by glucocorticoid induction of β-adrenergic receptor expression. Agonist-induced down-regulation of β-adrenergic receptor mRNA via message destabilization best highlights post-transcriptional control. Examples of post-translational control of transmembrane signalling elements include protein phosphorylation, thiol–disulphide exchange, and altered rates of protein degradation. Simultaneous analysis of physiological regulation at the levels of the gene, mRNA, and protein provide new opportunities for understanding how information processing extends from the plasma membrane to the genome.

Introduction

Understanding how information is transduced across biological membranes remains a major goal in biology. A variety of motifs by which cells respond to external signals have been identified in recent years. Prominent among these motifs is G-protein-mediated transmembrane signalling. This motif requires the interaction of members of three distinct classes of membrane proteins: receptors, G-proteins, and effector units (Gilman, 1987; Birnbaumer *et al.*,

1987; Lochrie & Simon, 1988; Milligan, 1988). Receptors discriminate among incoming signals, which may be a ligand such as adrenaline for the β-adrenergic receptor or, as in the case of vision, photons captured by rhodopsin. Agonist-bound receptors (or photobleached pigments) propagate signals to effector units indirectly, via interaction with and activation of G-proteins. Exchange of GTP for GDP by the α-subunit of G-proteins is facilitated by activated receptors. Activated GTP-liganded G-proteins, in turn, alter the activity of specific effector molecules, that may be enzymes or ion channels.

In photoexcitation, bleached rhodopsin activates a retinal-specific G-protein, G_t, which subsequently activates a cyclic GMP-specific phosphodiesterase in the vertebrate rod (Stryer & Bourne, 1986). The fall in cyclic GMP results in a change in membrane conductance characteristic of visual excitation. Catecholamine hormone action provides a second example for consideration. Catecholamine binding to β-adrenergic receptors results in the activation of the stimulatory G-protein, G_s, which then increases adenylyl-cyclase activity and raises intracellular cyclic AMP levels. Via other classes of receptors, catecholamines can act to decrease adenylyl cyclase activity, operating via members of a family of G-proteins, termed G_i, that appear capable of mediating the 'inhibitory' control of adenylyl cyclase. The population of G-protein-linked receptors continues to expand, as do the families of G-proteins (Strathmann *et al.*, 1989) and effectors (Brown *et al.*, 1989). Thus, endocrine and sensory physiology are replete with examples in which transmembrane signalling operates via G-protein-linked pathways.

Although much information has accumulated on the structure and function of elements of G-protein-linked pathways, much less is known about the cell biology and physiological regulation of these systems. The focus of this article will be a discussion of the regulation of transmembrane signalling, employing the hormone-sensitive adenylyl cyclase pathway as a model. In particular, recent advances in our understanding of the biology of β-adrenergic receptors will be used to highlight mechanisms by which transmembrane signalling elements may be regulated. Regulation at three levels: (i) transcriptional, (ii) post-transcriptional, and (iii) post-translational, will be addressed.

Transcriptional Regulation

Physiological regulation of G-protein-linked transmembrane signalling elements can be approached as two basic forms, up-regulation, in which the steady-state expression of an element increases, and down-regulation, in which the steady-state level declines (Malbon *et al.*, 1988). Glucocorticoids enhance catecholamine-stimulated responses propagated via β-adrenergic receptors (Davies & Lefkowitz, 1984). Glucocorticoids have been shown to increase expression of the most proximal element of the hormone-sensitive adenylyl cyclase system, the β-adrenergic receptor itself. Adrenalectomy is associated with a reduction in both parameters, an effect rectified by administration of glucocorticoids. Similarly, testosterone administration has been shown to increase and orchidectomy to depress β-adrenergic receptor levels in the prostate (Collins *et al.*, 1988*a*). Although the phenomenology of steroid-induced increases in receptor-mediated responses and receptor number has been known for many years, only recently has the molecular basis of this regulation been revealed.

Our understanding of steroid-induced up-regulation of β-adrenergic receptors advanced when mRNA levels for these low-abundance membrane-proteins were analysed. Using Northern (RNA blot) analysis, Collins *et al.* (1988*a*) demonstrated that receptor mRNA levels increased in a hamster vas deferens cell line (DDT_1 MF-2) exposed to glucocorticoid in culture. Quantification of mRNA levels, made possible by DNA excess solution hybridization, revealed the steady-state levels of β_2-adrenergic receptor mRNA to be $\sim$0.65 amol (10^{-18} mol)/μg of total cellular RNA in unstimulated cells (Hadcock & Malbon, 1988*a*). Receptor mRNA levels increased nearly 3-fold within 2 h of challenge with glucocorticoid (dexamethasone). This increase in mRNA was transient, declining in cells exposed to a maximal concentration of steroid for 4–24 h to a new steady-state only twice that of the control cells (Hadcock & Malbon, 1988*b*).

The basis for the increase in steady-state levels of receptor mRNA in response to steroid-induced up-regulation was clarified by several lines of experimentation. In the presence of actinomycin D, a potent inhibitor of RNA synthesis, glucocorticoid-promoted increases in receptor mRNA were not observed (Hadcock & Malbon, 1988*b*). The half-life of the β_2-adrenergic receptor mRNA was determined to be about 12 h and found to remain essentially unchanged in cells exposed to steroid. Dexamethasone stimulated a 4–5-fold increase in the rate of transcription of the gene for the β_2-adrenergic receptor in cells challenged with the glucocorticoid for 2 h (Collins *et al.*, 1988; Hadcock *et al.*, 1989*b*). By 72 h, glucocorticoids increase the expression of β_2-adrenergic receptor by 2-fold, as evidenced by indirect immunofluorescence with anti-receptor antibodies or by radioligand binding (Hadcock & Malbon, 1988*b*).

Does this increase in transcription and receptor expression explain the enhanced response typically observed in glucocorticoid-treated cells or animals? To approach this fundamental question, Chinese hamster ovary (CHO) cells were stably transfected with an expression vector that harbours the cDNA for the hamster β_2-receptor under the control of the Simian Virus 40 early promoter (George *et al.*, 1988). Clones that stably express 200–2 000 000 receptors per cell were examined for cyclic AMP accumulation in response to the β-adrenergic agonist (−)-isoprenaline. Both the maximal response and the sensitivity of the cyclic AMP response increased proportionally with receptor expression, providing compelling evidence that the enhanced physiological response promoted by glucocorticoids can be explained, in fact, by the steroid-induced expression of receptor.

At the level of the β_2-adrenergic receptor gene, several putative glucocorticoid-response elements (GRE) have been identified (Chung *et al.*, 1987; Emorine *et al.*, 1987; Kobilka *et al.*, 1987). These putative GREs are located in the 5′-untranslated, 3′-untranslated and open-reading-frame domains of the gene. Recent evidence implicates a GRE in the 5′-untranslated region as obligate for glucocorticoid sensitivity (Malbon & Hadcock, 1988).

The presence of GREs in the open-reading frame and 3′-untranslated regions failed to confer steroid sensitivity to the gene.

To what extent does transcriptional activation of the β_2-adrenergic receptor gene by glucocorticoids represent a general paradigm for physiological regulation of other elements of G-protein-linked transmembrane signalling? Although only speculation, it is likely that the regulation by glucocortoids of $G_i2\alpha$ and the G_β-subunits common to G_s, G_i, and G_o may also include a transcriptional component (Ros *et al.*, 1989*a,b*). It is well known that steroid hormone receptors and thyroid hormone receptors can exert both positive and negative effects on transcription. Thyroid hormones, too, have been shown to regulate the expression of $G_s\alpha$- and G_β-subunits (Ros *et al.*, 1988; Rapiejko *et al.*, 1989). Unlike glucocorticoids, thyroid hormones decrease G_β-subunit expression, while hypothyroidism increases expression. Thus transcriptional activation may explain many facets of the physiological regulation of transmembrane signalling. Transcriptional repression (Levine & Manley, 1989), too, may participate in the regulation.

Post-Transcriptional Regulation

Adaptation, the decline in sensitivity to a stimulus following chronic challenge, is nearly universally found in biology. In vision, adaptation follows exposure to light, whereas in olfaction adaptation follows exposure to pleasant and noxious odorants, alike. In addition to sensory physiology, endocrine physiology is replete with examples of adaptation to chronic exposure to hormones, a process generally termed 'desensitization' (Sibley & Lefkowitz, 1985). Two forms of desensitization, homologous and heterologous, have been described. If chronic challenge by a stimulus adapts only the response to that stimulus, the desensitization is termed homologous. If the challenge adapts responses to stimuli other than itself, the desensitization is termed heterologous.

Agonists promote not only a desensitization of the response in the short term, but also a down-regulation of receptor over the longer term (Mahan *et al.*, 1987). For G-protein-linked receptors, agonist-promoted down-regulation is commonly observed. Only recently has insight into the underlying mechanisms of down-regulation been revealed. Chronic treatment of cells with β-adrenergic agonists results in a sharp decline in receptor binding. Neve & Molinoff (1986) demonstrated that the rate of recovery of β-adrenergic receptors was slower in cells for which receptor had been down-regulated by agonist than in cells treated with reagents that chemically inactivate receptors. Newly established steady-state levels of β-receptors in cells previously down-regulated with agonist were significantly less than those of control cells. Based upon this and other evidence, it appeared likely that down-regulation of receptor by agonist might be exerted, in part, at the level of receptor mRNA.

Hadcock & Malbon (1988*a*) first showed that challenging cells with agonist resulted in a down-regulation of β_2-adrenergic receptor mRNA levels. The response was dose dependent with respect to agonist, time-dependent, reversible and displayed the pharmacology expected of a β_2-receptor-mediated process. The dose response for agonist-promoted down-regulation agreed well with that for agonist-stimulated cyclic AMP accumulation in those cells. Interestingly, although stimulating cyclic AMP accumulation to the same extent as the diterpene forskolin, β-agonist promoted a greater decline in receptor mRNA levels. These data suggested that some additional mediator(s) other than cyclic AMP may participate in agonist-promoted down-regulation of receptor message.

The role of cyclic AMP in down-regulation was further analysed in a series of well-defined variants of the S49 mouse lymphoma cell lines (Hadcock *et al.*, 1989*a*). Variants with mutations in $G_s\alpha$ displayed interruption of receptor–G_s coupling (*unc*), no $G_s\alpha$ mRNA or expression (cyc^-), and interruption of G_s-adenylyl cyclase coupling (*H21A*). Forskolin activated adenylyl cyclase in wild-type and $G_s\alpha$-mutants alike, resulting in an elevation of cyclic AMP accumulation and a down-regulation of receptor mRNA. Isoprenaline, in contrast, failed to stimulate cyclic AMP accumulation in $G_s\alpha$-mutant cell lines. Most interesting was the observation that the *H21A* mutant cells displayed no change in cyclic AMP accumulation, but a down-regulation of receptor mRNA. But what then was the role of cyclic AMP? To pursue this question, we examined an S49 variant (kin^-) with a transdominant mutation in the catalytic domain of the cyclic AMP-dependent protein kinase (A kinase). Both forskolin and isoprenaline stimulated cyclic AMP accumulation, but failed to down-regulate receptor mRNA levels and expression in kin^- cells. Thus basal cyclic AMP levels and protein kinase A activity appear obligate for the down-regulation. The study of the *H21A* mutant identifies an additional pathway to down-regulation that presumably still requires protein kinase A activity. These results solidified the role of cyclic AMP in agonist-promoted down-regulation of receptor mRNA, but failed to provide the mechanism(s).

The identification of a cyclic AMP response element (CRE) in the β_2-adrenergic receptor gene raised the intriguing possibility that transcriptional repression was responsible for the decline in receptor mRNA in agonist-treated cells (Hadcock & Malbon, 1988*a*). This possibility was addressed by two approaches. Using nuclear run-on assays to examine transcription, Collins *et al.* (1989*a*) found an increase, rather than a decrease, in transcription in agonist-treated cells. This increase in transcription was only transient and of modest magnitude (0.7-fold over basal). Analysis at later times, by others, revealed no significant change in the relative rates of transcription for the β_2-adrenergic receptor gene (Hadcock *et al.*, 1989*b*).

Analysis of the half-life of receptor mRNA provided a key observation. Previously, the half-life of the β_2-receptor mRNA was shown to be about 12 h and not influenced by glucocorticoid treatment. Measurements performed on cells treated with agonist revealed that mRNA half-life declined from 12 to 5 h. This destabilization of mRNA appears to be the basis for agonist-promoted down-regulation of receptor message. Thus, the ability of cells to adapt to chronic activation of the stimulatory pathway of adenylyl cyclase involves a post-transcriptional mechanism, the destabilization of mRNA of the receptor that mediated the response. Since agents that elevate intracellular cyclic AMP also down-regulate receptor mRNA and expression, it appears that destabilization of mRNA is at least one mechanism of heterologous down-regulation.

Post-Translational Regulation

Desensitization of G-protein-linked receptors in general, and β-adrenergic receptors in particular, has focused much attention upon regulation of transmembrane signalling by post-translational mechanisms. Two phenomena, receptor sequestration and protein phosphorylation, have gained wide acceptance as candidates for desensitization mechanisms (Sibley & Lefkowitz, 1985). Using indirect immunofluorescence and anti-receptor antibodies, Wang *et al.* (1989*a,b*) have examined β-adrenergic receptors of fixed, intact cells that were pretreated with and without β-adrenergic agonist. The results of this work demonstrated receptor immunoreactivity in punctate staining largely confined to the cell membrane. The amount and general pattern of staining observed in control and agonist-treated cells were equivalent, suggesting that the bulk of the receptors were not shielded from antibody (i.e. internalized) by challenge with agonist. Biochemical studies clearly demonstrate a loss of radioligand binding in response to challenge with agonist, with no major alteration of cellular complement receptor or receptor integrity (M_r). Receptor may be sequestered from G-protein and effector units during desensitization. However, there is little data to suggest that the receptors have been sequestered away from the cell membrane during desensitization (Wang *et al.*, 1989*b*).

Protein phosphorylation is a prominent mechanism of physiological regulation at the post-translational level (Sibley *et al.*, 1987). G-protein-linked receptors are substrates for phosphorylation by protein kinases A and C, as well as a β-adrenergic-receptor-specific kinase, βARK (Benovic *et al.*, 1986). Protein kinase A has been shown to be required for short-term agonist-promoted desensitization (Clark *et al.*, 1988) and long-term agonist-promoted down-regulation of receptor mRNA and expression (Hadcock *et al.*, 1989*a*). Evidence has accumulated that activation of protein kinase C by phorbol esters, too, results in phosphorylation of the β-adrenergic receptor and desensitization (Kelleher *et al.*, 1984). βARK selectively phosphorylates only the agonist-occupied form of the receptor. Unlike protein kinase A, βARK appears to phosphorylate the receptor only at relatively high concentrations of agonist (Benovic *et al.*, 1989). Protein kinase A has been shown to mediate desensitization at lower (physiological) levels of agonist. In spite of the demonstration that G-protein-linked receptors are substrates for phosphorylation, the precise manner in which the phosphorylation alters receptor function remains unresolved. Although changes in receptor–G_s coupling have been reported for phosphorylated receptor co-reconstituted with purified G_s, these changes are modest and fail to account for the loss of receptor function observed in desensitization (Benovic *et al.*, 1985). Perhaps some additional protein, like arrestin, is required for the expression of the uncoupling between the phosphorylated receptor and G_s (Benovic *et al.*, 1987)? The temporal relationship between net phosphorylation of the receptor and agonist-promoted desensitization has not been detailed.

We have been intrigued by the images of β-adrenergic receptors obtained from indirect immunofluorescence staining of intact, fixed cells (Wang *et al.*, 1989*a,b*). The punctate staining patterns observed may indicate macromolecular organization of β-adrenergic receptors in clusters, much like those observed for the nicotinic acetylcholine receptor. Recently it has been shown that upon desensitization of the nicotinic receptor in chick muscle cells, receptor is not internalized, but rather undergoes redistribution from large patches to smaller patches (Ross *et al.*, 1988). Perhaps G-protein-linked receptors, too, undergo redistribution away from the other elements of the system by a process that may involve phosphorylation. Recent observations from our laboratory support this notion, at least for β-adrenergic receptors in agonist-treated rat osteosarcoma cells.

There exist compelling data to implicate intramolecular disulphide bridge–thiol exchange in the structure and function of G-protein-linked receptors (Malbon *et al.*, 1987). The ligand-binding capacity of β-adrenergic receptors is destabilized by treatment with thiols. Many G-protein-linked receptors isolated to date display a thiol-dependent change in their migration on SDS/PAGE. Under non-reducing conditions, the β-adrenergic receptor migrates with M_r 55 000, while the fully reduced and alkylated receptor migrates with M_r 65 000 (Moxham & Malbon, 1985). Pedersen & Ross (1985) first demonstrated that thiol-treated β-adrenergic receptors can activate G_s in a manner similar to agonist-bound receptor. Furthermore, the native receptor in the membrane was found to be the 55 000-M_r species (Moxham *et al.*, 1988). These data implicate disulphide–thiol exchange in the structure and function of the β-adrenergic receptor, and perhaps other members of the G-protein-linked receptor family.

G-proteins are targets for several post-translational modifications, some of which are known to modify the functional status of the molecule. Bacterial toxins catalyse the specific ADP-ribosylation of the α-subunits of many of the G-proteins that mediate transmembrane signalling (Milligan, 1988). $G_s\alpha$-subunits are subject to ADP-ribosylation (at the expense of NAD^+) by cholera toxin, whereas members of the $G_i\alpha$ family and $G_o\alpha$ are targets for ADP-ribosylation by pertussis toxin. The functional outcome of this mono-ADP-ribosylation by the toxins is quite distinct. Cholera toxin treatment results in activation of $G_s\alpha$. Pertussis toxin treatment, in contrast, inactivates its G-protein substrates. One G-protein, G_t, found in high abundance in vertebrate retinal rods is a substrate for both cholera and pertussis toxins (Stryer & Bourne, 1986). Although endogenous mono-ADP-ribosyltransferases have been identified in cellular extracts (Vaughan & Moss, 1982), it is not clear to what extent these activities regulate the function of the cellular complement of G-proteins. *N*-linked myristoylation, too, occurs post-translationally for some G-proteins (Freissmuth *et al.*, 1989). Palmitoylation of G-protein-linked receptors has been reported also (Ovchinnikov *et al.*, 1988; O'Dowd *et al.*, 1989). The functional consequence of this modification and the extent to which fatty acylation provides a basis for regulation of function are not known.

Perhaps the least studied of post-translational regulation of G-protein-linked transmembrane signalling elements is that at the level of protein turnover. A possible role of protein turnover was highlighted by several recent observations. It has been observed that once ADP-ribosylated by the action of cholera toxin, $G_s\alpha$ may be degraded at an accelerated rate (Chang & Bourne, 1989; Milligan *et al.*, 1989). In addition, pertussis toxin treatment of 3T3-L1 and S49 mouse lymphoma cells resulted in no apparent change in the steady-state levels of $G_i\alpha$s and $G_o\alpha$, but a sharp decline in G_β-subunits (Watkins *et al.*, 1989). Similar changes in the level of G-proteins were observed in fat cells acutely prepared from animals intoxicated with pertussis toxin *in vivo*. This decline in G_β levels after toxin treatment was not accompanied by a dramatic decline in the mRNA for G_β. Thus it seems likely that enhanced turnover of the G_β-subunits may best explain the phenomenon. Indirect approaches to the question of receptor turnover have been reported. However, there exists very little data on the turnover of G-protein-linked receptors.

Summary

Fig. 1 depicts the various mechanisms identified to date that participate in the regulation of G-protein-linked transmembrane signalling elements. The up-regulation of β-adrenergic receptors by glucocorticoids exemplifies regulation at the level of transcription, in this case a steroid-induced activation. Short-term agonist treatment also leads to an increase in mRNA for the β_2-adrenergic receptor, though transient. Post-translational regulation, altering the stability of mRNA, appears to be a mechanism central to down-regulation of β-adrenergic receptors, and perhaps other G-protein-linked receptors. Agonist-promoted destabilization of receptor mRNA is a key facet

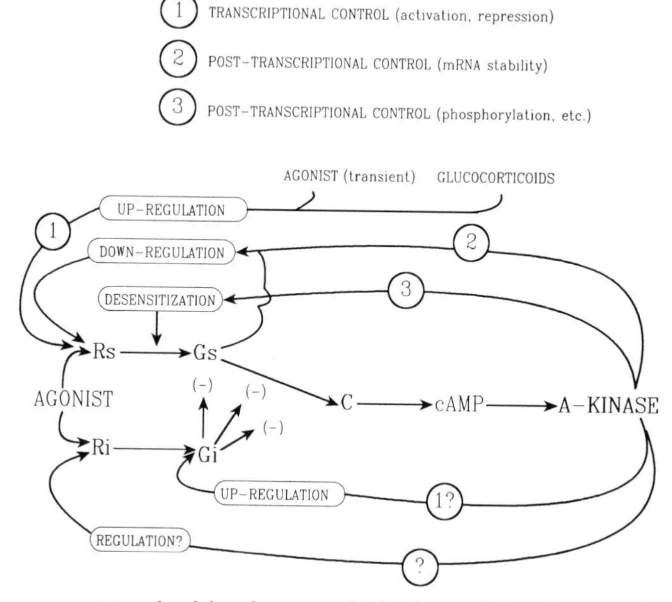

Fig. 1. *Hormone-sensitive adenylyl cyclase: a model for the regulation of transmembrane signalling at the transcriptional, post-transcriptional, and post-translational levels*

of down-regulation of β-adrenergic receptor expression. Protein phosphorylation provides a prominent example of post-translational control mechanisms. Protein kinases have been implicated in desensitization of the stimulatory pathway of adenylyl cyclase. Protein kinase A has been shown to mediate heterologous desensitization, whereas βARK appears to participate in homologous desensitization. G-proteins are substrates for several post-translational modifications. In the case of bacterial toxin-catalysed ADP-ribosylation, this covalent modification alters the functional status of the molecule. Finally, for many regulatory phenomena in the biology of transmembrane signalling, insights into underlying mechanisms are lacking. Activation of the stimulatory pathway ($R_s \rangle\rangle\rangle G_s \rangle\rangle\rangle C \rangle\rangle\rangle$, etc.) results in an increase in the expression of elements in the opposing, inhibitory pathway. To what extent such 'cross-talk' involves transcriptional, post-transcriptional, and post-translational mechanisms remains to be established. With a greater appreciation of the diversity of mechanisms available for regulating the expression of transmembrane signalling elements, we can begin to approach the more complex sphere of how information via multiple G-protein-linked pathways is integrated and processed.

This work was supported by U.S. Public Health Services grants DK30111, DK25410, and K04-AM00786 from the National Institute of Diabetes, Digestive, and Kidney Diseases, N.I.H.

References

Benovic, J. L., Pike, L. J., Cerione, R. A., Staniszewski, C., Yoshimasa, T., Codina, J., Caron, M. G. & Lefkowitz, R. J. (1985) *J. Biol. Chem.* **260**, 7094–7101

Benovic, J. L., Strasser, R. H., Caron, M. G. & Lefkowitz, R. J. (1986) *Proc. Natl. Acad. Sci. U.S.A.* **83**, 2797–2801

Benovic, J. L., Kuhn, H., Weyand, I., Codina, J., Caron, M. G. & Lefkowitz, R. J. (1987) *Proc. Natl. Acad. Sci. U.S.A.* **84**, 8879–8882

Benovic, J. L., DeBlasi, A., Stone, W. C., Caron, M. G. & Lefkowitz, R. J. (1989) *Science* **246**, 235–240

Birnbaumer, L., Codina, J., Mattera, M., Yatani, A., Scherer, N., Toro, M.-Y. & Brown, A. B. (1987) *Kidney Int.* **32** (Suppl. 23), S14–S37

Brown, A. M., Yatani, A., Codina, J. & Birnbaumer, L. (1989) *Am. J. Hypertens.* **2**, 124–127

Chang, F.-H. & Bourne, H. R. (1989) *J. Biol. Chem.* **264**, 5352–5357

Chung, F.-Z., Ulrich Lentes, K., Gocayne, J., Fitzgerald, M., Robinson, D., Kervalage, A. R., Fraser, C. M. & Venter, J. C. (1987) *FEBS Lett.* **211**, 200–206

Clark, R. B., Kunkel, M. W., Friedman, J., Goka, T. J. & Johnson, J. A. (1988) *Proc. Natl. Acad. Sci. U.S.A.* **85**, 1442–1446

Collins, S., Caron, M. G. & Lefkowitz, R. J. (1988*a*) *J. Biol. Chem.* **263**, 9067–9070

Collins, S., Quarmby, V. E., French, F. S., Lefkowitz, R. J. & Caron, M. G. (1988*b*) *FEBS Lett.* **233**, 173–176

Collins, S., Bouvier, M., Bolanowski, M. A., Caron, M. G. & Lefkowitz, R. J. (1989) *Proc. Natl. Acad. Sci. U.S.A.* **86**, 4853–4857

Davies, A. O. & Lefkowitz, R. J. (1984) *Annu. Rev. Physiol.* **46**, 119–130

Emorine, L. J., Marulo, S., Delavier-Klutchko, C., Kaveri, S. V., Durieu-Trautman, O. & Strosberg, A. D. (1987) *Proc. Natl. Acad. Sci. U.S.A.* **84**, 6995–6999

Freissmuth, M., Casey, P. J. & Gilman, A. G. (1989) *FASEB J.* **3**, 2125–2131

George, S. T., Berrios, M., Hadcock, J. R., Wang, H.-Y. & Malbon, C. C. (1988) *Biochem. Biophys. Res. Commun.* **150**, 665–672

Gilman, A. G. (1987) *Annu. Rev. Biochem.* **56**, 615–649

Hadcock, J. R. & Malbon, C. C. (1988*a*) *Proc. Natl. Acad. Sci. U.S.A.* **85**, 5021–5025

Hadcock, J. R. & Malbon, C. C. (1988*b*) *Proc. Natl. Acad. Sci. U.S.A.* **85**, 8415–8419

Hadcock, J. R., Ros, M. & Malbon, C. C. (1989*a*) *J. Biol. Chem.* **264**, 13956–13961
Hadcock, J. R., Wang, H.-Y. & Malbon, C. C. (1989*b*) *J. Biol. Chem.* **264**, 19928–19933
Kelleher, D. J., Pessin, J. E., Ruoho, A. E. & Johnson, G. L. (1984) *Proc. Natl. Acad. Sci. U.S.A.* **81**, 4316–4320
Kobilka, B. K., Frielle, T., Dohlman, H. G., Bolanowski, M. A., Dixon, R. A. F., Keller, P., Caron, M. G. & Lefkowitz, R. J. (1987) *J. Biol. Chem.* **262**, 7321–7327
Levine, M. & Manley, J. L. (1989) *Cell (Cambridge, Mass.)* **59**, 405–408
Lochrie, M. A. & Simon, M. I. (1988) *Biochemistry* **27**, 4957–4965
Mahan, L. C., McKernan, R. M. & Insel, P. A. (1987) *Annu. Rev. Pharmacol. Toxicol.* **27**, 215–235
Malbon, C. C. & Hadcock, J. R. (1988) *Biochem. Biophys. Res. Commun.* **154**, 676–681
Malbon, C. C., George, S. T. & Moxham, C. P. (1987) *Trends Biochem. Sci.* **12**, 172–176
Malbon, C. C., Rapiejko, P. J. & Watkins, D. C. (1988) *Trends Pharmacol. Sci.* **9**, 33–36
Milligan, G. (1988) *Biochem. J.* **255**, 1–13
Milligan, G., Unson, C. G. & Wakelam, M. J. O. (1989) *Biochem. J.* **262**, 643–649
Moxham, C. P. & Malbon, C. C. (1985) *Biochemistry* **24**, 6072–6077
Moxham, C. P., Ross, E. M., George, S. T., Brandwein, H. & Malbon, C. C. (1988) *Mol. Pharmacol.* **33**, 486–492
Neve, K. A. & Molinoff, P. B. (1986) *Mol. Pharmacol.* **30**, 104–111
O'Dowd, B. F., Hnatowich, M., Caron, M. G., Lefkowitz, R. J. & Bouvier, M. (1989) *J. Biol. Chem.* **264**, 7564–7569
Ovchinnikov, Y. A., Abdulaev, N. G. & Bogachuk, A. S. (1988) *FEBS Lett.* **230**, 1–5
Pedersen, S. E. & Ross, E. M. (1985) *J. Biol. Chem.* **260**, 14150–14157
Rapiejko, P. J., Watkins, D. C., Ros, M. & Malbon, C. C. (1989) *J. Biol. Chem.* **264**, 16183–16189
Ros, M., Northup, J. K. & Malbon, C. C. (1988) *J. Biol. Chem.* **263**, 4362–4368
Ros, M., Northup, J. K. & Malbon, C. C. (1989*a*) *Biochem. J.* **257**, 737–744
Ros, M., Watkins, D. C., Rapiejko, P. J. & Malbon, C. C. (1989*b*) *Biochem. J.* **260**, 271–275
Ross, A., Rapuano, M. & Prives, J. (1988) *J. Cell Biol.* **107**, 1139–1145
Sibley, D. R. & Lefkowitz, R. J. (1985) *Nature (London)* **317**, 124–129
Sibley, D. R., Benovic, J. L., Caron, M. G. & Lefkowitz, R. J. (1987) *Cell (Cambridge, Mass.)* **48**, 913–922
Strathmann, M., Wilkie, T. M. & Simon, M. I. (1989) *Proc. Natl. Acad. Sci. U.S.A.* **86**, 7407–7409
Stryer, L. & Bourne, J. R. (1986) *Annu. Rev. Cell Biol.* **2**, 391–419
Vaughan, M. & Moss, J. (1982) *Curr. Top. Cell. Regul.* **20**, 205–246
Wang, H.-Y., Berrios, M. & Malbon, C. C. (1989*a*) *Biochem. J.* **263**, 519–532
Wang, H.-Y., Berrios, M. & Malbon, C. C. (1989*b*) *Biochem. J.* **263**, 533–538
Watkins, D. C., Northup, J. K. & Malbon, C. C. (1989) *J. Biol. Chem.* **264**, 4186–4194

Biochem. Soc. Symp. **56**, 165–170
Printed in Great Britain

GTPase-Inhibiting Mutations Activate the α-Chain of G_s in Human Tumours

LUCIA VALLAR

Department of Pharmacology, C.N.R. Center of Cytopharmacology, Scientific Institute San Raffaele, University of Milan, Milan 20132, Italy

Synopsis

Several trophic hormones transmit their signals by coupling to G-proteins. A subset of human growth hormone (GH)-secreting pituitary tumours, characterized by elevated GH secretion, cyclic AMP levels, and adenylyl cyclase activity, carries mutations in the gene that encodes the α-chain of G_s. By substitution of a single amino acid (Arg-201 or Gln-227), these mutations inhibit the intrinsic GTPase activity of α_s, thus stabilizing the protein in its active conformation. The discovery of mutant α_s proteins in human tumours suggests that the α_s gene can be converted into an oncogene, called *gsp*, in cells that proliferate in response to cyclic AMP. Most likely, oncogenes can result from similar mutations in other G-protein α-chain genes.

Introduction

Heterotrimeric G-proteins couple cell-surface receptors for hormones and neurotransmitters to effector molecules, either enzymes or ion channels, that control the accumulation of second messengers [1–3]. The structural and functional properties of signalling G-proteins have been characterized in great detail [1–3]. G-proteins consist of three polypeptides: a guanine-nucleotide-binding α-chain, a β-chain, and a γ-chain. An individual G-protein cycles between an inactive GDP-bound state and an active GTP-bound state. When GDP is bound, the α-chain associates with high affinity to βγ; the rate of dissociation of GDP from the complex is extremely slow. Receptor activation promotes the dissociation of GDP, thereby accelerating the binding of GTP. The conformational change induced by the binding of GTP to α results in a marked decrease in affinity for the βγ-complex. α_{GTP}, released from βγ, can interact with and turn on the target effector molecules(s). Activation is then terminated by the built-in GTPase activity of the α-chain, which hydrolyses the bound nucleotide to return to its inactive state.

The G-protein-coupling mechanism is employed by a variety of extracellular signals to generate intracellular second messengers that regulate multiple functions, including cell growth [1–5]. The role of G-proteins in mediating mitogenic signals predicts that oncogenic mutations should be found in the genes that encode their α-chains. Recent work revealing that indeed certain human

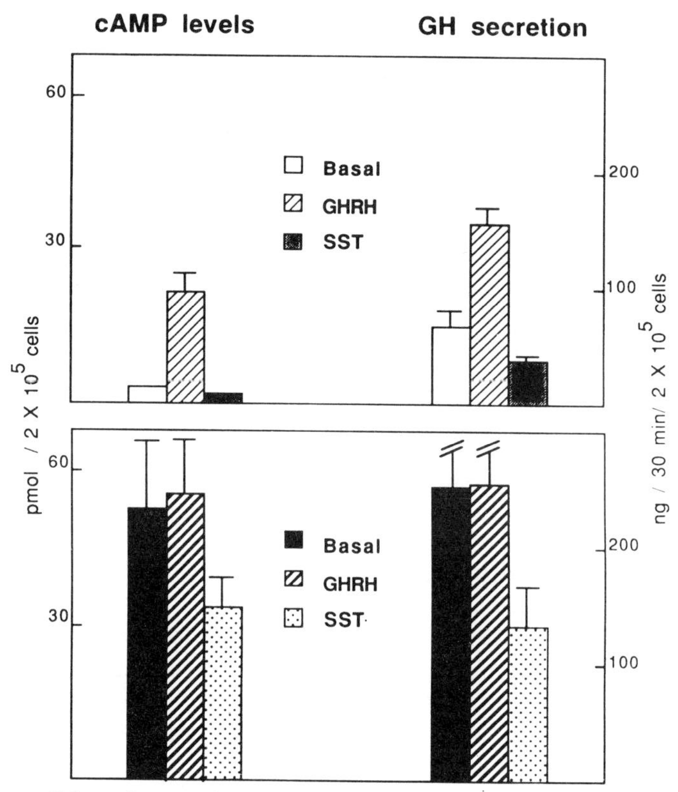

Fig. 1. *Intracellular cyclic AMP levels and GH secretion in human GH-secreting pituitary tumours*

The upper panel shows cyclic AMP (cAMP) levels and GH secretion under basal conditions and in the presence of 10 nM-GHRH or 100 nM-somatostatin (SST) in short-term cultured cells from group 1 tumours ($n = 17$). The results obtained with cells from group 2 tumours ($n = 8$) are illustrated in the lower panel.

tumours carry somatic mutations in the G-protein α-chain genes will be summarized here. These mutations cause constitutive activation of α_s, the α-subunit of the stimulatory regulator of adenylyl cyclase, by inhibiting its intrinsic GTPase activity.

Constitutive Activation of G_s in Human Pituitary Tumours

Cyclic AMP acts as a positive intracellular signal for cell proliferation in several endocrine tissues, such as the thyroid and some cells of the pituitary, and a few other cell types [5,6]. GH-releasing hormone (GHRH) uses cyclic AMP as a second messenger to stimulate GH secretion and proliferation of normal pituitary somatotrophs [6]. The biochemical characterization of human GH-secreting pituitary adenomas, the tumours which arise from somatotroph cells, revealed that in a large percentage of cases the adenylyl cyclase–cyclic AMP pathway is constitutively activated in the tumour cells, thus bypassing the normal requirement for GHRH [7]. Two distinct groups of tumours were identified: group 1 tumours, like normal somatotroph cells, had low 'basal' cyclic AMP levels and GH secretion, which markedly increased in the presence of GHRH; in contrast, in group 2 tumours 'basal' intracellular cyclic AMP concentration and secretory activity were elevated and responded poorly or not at all to further stimulation with the neurohormone (Fig. 1). The patterns of adenylyl cyclase responsiveness in membranes of the two sets of tumours are summarized in Fig. 2. Group 2 tumours showed a marked elevation of basal adenylyl cyclase activity when compared with group 1 tumours. In the latter group of tumours GTP, GTP plus GHRH, the stable GTP analogue, guanosine 5′-[β,γ-imido]triphosphate (Gpp[NH]p), and fluoride, all stimulated adenylyl cyclase activity. In group 2 tumours, GTP, Gpp(NH)p, and fluoride either had no effect or inhibited the enzyme activity. GHRH stimulation, in the presence of GTP, was barely detectable. Forskolin-stimulated adenylyl cyclase was much higher in group 2 than in group 1. Both sets of tumours showed normal responsiveness to the inhibitory receptor agonist, somatostatin. These latter results were in keeping with the ability of this neurohormone to inhibit cyclic AMP levels and GH secretion in intact tumour cells (Fig. 1).

The hormone-independent induction of cyclic AMP synthesis in group 2 tumours could result from the autonomous activation of any one of the three well-characterized components of the membrane adenylyl cyclase system: the hormone receptor, the adenylyl cyclase catalytic unit or the stimulatory

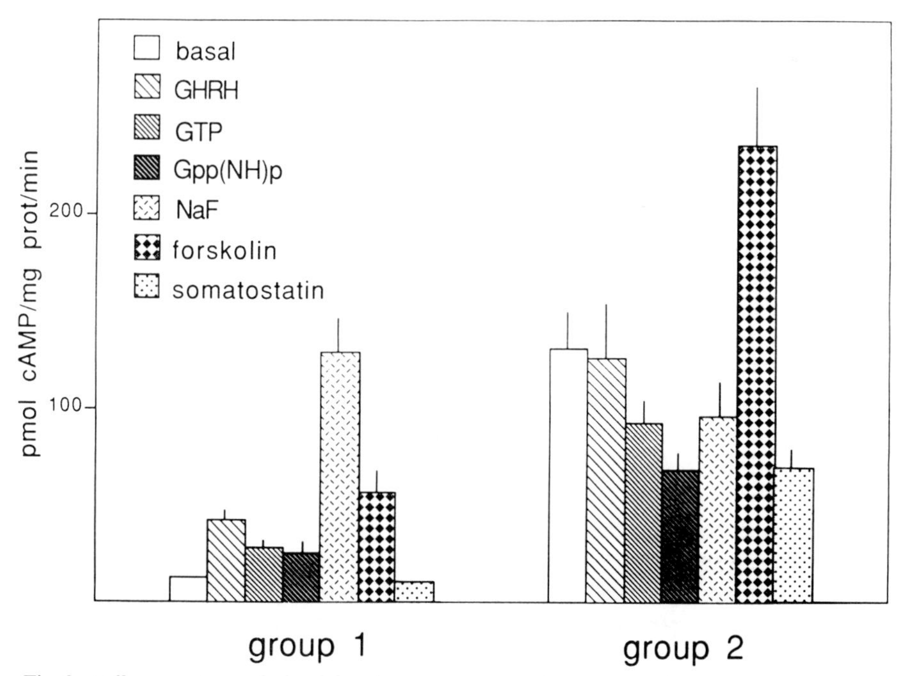

Fig. 2. *Different patterns of adenylyl cyclase responsiveness in group 1 and group 2 pituitary tumours*

Adenylyl cyclase activity was measured in membranes from group 1 tumours (n = 6–48) and group 2 tumours (n = 6–20) as described in [7]. The concentrations of the various test agents were: GTP, 10 μM; GHRH (in the presence of GTP), 1 μM; Gpp(NH)p, 10 μM; NaF, 10 mM; forskolin 100 μM; somatostatin (in the presence of GTP), 10 μM.

G-protein, G_s. In all the cases characterized so far, group 2 tumour cells were found to express a constitutively active form of G_s. Indeed, when G_s extracts from group 2 tumours were tested for their ability to reconstitute the membrane adenylyl cyclase activity of α_s-deficient S49 *cyc*$^-$ cells, they were able to reproduce the constitutive activation observed in the original tumour membranes [7,8].

GTPase-Inhibiting Mutations Activate the α-Chain of G_s in Human Pituitary Tumours

The biochemical phenotype of group 2 tumours was entirely consistent with the possibility of oncogenic mutations in the G_s α-chain gene. The cloning and sequencing of α_s cDNAs from four group 2 tumours revealed mutations in all cases [8]. Mutations in three tumours replaced Arg-201 with either Cys or His (using the single letter code for amino acids these mutations are designed R201C and R201H, respectively). In a fourth tumour Gln-227 was replaced by Arg (Q227R). Genomic DNA from each tumour showed the same mutation detected in the corresponding cDNA as well as wild-type sequence at the same position, suggesting the presence of both a normal and a mutant α_s allele. In keeping with the expectation of a somatic origin of the α_s mutations, genomic DNA from peripheral blood cells of two patients with group 2 tumours contained only wild-type sequence [8]. Genomic DNA from group 1 tumours always showed only wild-type sequence at codons 201 and 227 [8]. To determine whether the mutations detected in group 2 tumours cause constitutive activation of α_s, R201C, R201H and Q227R mutant rat α_s proteins were expressed in *cyc*$^-$ cells. Each of the three mutant proteins reproduced the adenylyl cyclase phenotype observed in group 2 tumours ([7,8] and Fig. 2).

In principle, the Gln-227 and Arg-201 mutations could activate G_s in three ways. Either type of mutation could: (i) induce an active conformational change that does not require the binding of guanine nucleotides; (ii) mimic the normal effect of the hormone–receptor complex, which activates α_s by increasing the rate at which GDP dissociates from its inactive form, thereby accelerating binding of GTP and attainment of the active state; (iii) inhibit the intrinsic GTPase of α_s, which serves as a timing device that turns off the active conformation of the protein [1–3]. This last mechanism turned out to account for α_s activation by mutations at both sites. The Q227R, R201C and R201H mutations inhibited the GTPase activity of α_s expressed in *cyc*$^-$ cells ([8] and Table 1).

Both amino acid residues replaced by the tumour mutations were already known or suspected to be important in α_s function. Gln-227 is located in the putative guanine-nucleotide-binding region of α_s, at a position corresponding to Gln-61 of p21^{ras} [9]; mutational replacement of Gln-61 in p21^{ras} by almost any other amino acid produces a protein that has a reduced ability to hydrolyse GTP and promotes malignant transformation [10]. Similarities in primary structure between short stretches of conserved amino acid sequence in G-protein α-chains and part of the guanine-nucleotide-binding domain of p21^{ras} had suggested that these α-chain regions form a GTP-binding domain

Table 1. *GTPase activity of wild-type and mutant* α_s

$K_{cat.GTP}$ values were estimated by an indirect method, as described in [8].

α_s	$K_{cat.GTP}$ (min^{-1})
Wild-type	4.1
Q227R	0.12
R201C	0.12
R201H	0.15

structurally and functionally similar to that of p21^{ras} [9,11]. Even before the Q227R mutation was identified, site-directed mutagenesis studies had shown that replacement of Gln-227 by Leu (Q227L) activates α_s by inhibiting its GTPase activity [12,13]. Thus, phenotypes produced by the deliberately designed Q227L mutation and by the Q227R mutation found in a human pituitary tumour are consistent with the hypothesis that G-protein α-chains and p21^{ras} share similar guanine-nucleotide-binding domains and use similar or identical mechanisms to hydrolyse GTP.

Arg-201 is the residue thought to be ADP-ribosylated by cholera toxin, a covalent modification that activates α_s by inhibiting its GTPase activity [1–3]. The similar effects of the tumour-associated R201C and R201H mutations suggest a specific role for the shape and location of the Arg-201 side-chain in normal GTP hydrolysis. Arg-201 is located in domain II of α_s, a region that has no obvious counterpart in p21^{ras} [9]. It is, therefore, reasonable to speculate that domain II may perform a function that is lacking in p21^{ras}. The intrinsic rate of GTP hydrolysis by G-proteins represents one such function. p21^{ras} hydrolyses GTP extremely slowly in the absence of a specific GTPase-activating protein (GAP) [14]; in contrast, the rate of GTP hydrolysis by α-chains is relatively fast and unaffected by binding of other proteins [12,13, 15]. It is tempting to speculate that domain II serves as an intrinsic, 'built-in' GAP for G-protein α-chains.

The screening of a large number of human GH-secreting pituitary tumours revealed that GTPase-inhibiting mutations at Arg-201 and Gln-227 account for the activation of α_s in the vast majority (and possibly all) of group 2 tumours. Among 16 tumours characterized by elevated adenylyl cyclase activity, 12 contained mutations in codon 201 (in 11 cases Arg to Cys and in one case Arg to His) and 4 contained mutations in codon 227 (Gln replaced by Arg in one case and by Leu in three cases) (J. Lyons, C. A. Landis, G. Harsh, L. Vallar, K. Grunewald, H. Feichtinger, Q. Duh, O. H. Clark, E. Kawasaky, H. R. Bourne & F. McCormick, unpublished work; E. Clementi, N. Malgaretti & R. Taramelli, unpublished work). As discussed above, the Q227L mutation detected in three tumours is already known to inhibit α_s GTPase activity [12,13].

The *gsp* Oncogene

The identification of GTPase-inhibiting mutations that activate α_s in human pituitary tumours led to the proposal that the α_s gene can be converted into an oncogene, designed *gsp* (for G_s-protein), in those cells which are programmed to proliferate in response to cyclic AMP [8]. The most stringent criterion for qualification of *gsp* as an oncogene is that deliberate expression of the mutant α_s proteins should reproducibly induce the formation of tumours. Although such evidence is not yet available, further support for the *gsp* oncogene hypothesis has been recently provided by the detection of mutations at codon 201 of α_s in other human tumours, namely an autonomously functioning thyroid adenoma (J. Lyons, C. A. Landis, G. Harsh, L. Vallar, K. Grunewald, H. Feichtinger, Q. Duh, O. H. Clark, E. Kawasaky, H. R. Bourne & F. McCormick, unpublished work).

The common mechanism for binding and hydrolysing GTP of all G-protein α-chains and their highly conserved primary structure in regions that correspond to the location of Arg-201 and Gln-227 in α_s, predict that other G-proteins that mediate proliferative signals can be converted into oncogenes by GTPase-inhibiting mutations. Indeed, the screening of a large number of different human tumours for mutations in the α_i2 gene revealed substitution of Arg-179 (which corresponds to Arg-201 in α_s) in tumours derived from the adrenal cortex and from ovarian granulosa tissue (J. Lyons, C. A. Landis, G. Harsh, L. Vallar, K. Grunewald, H. Feichtinger, Q. Duh, O. H. Clark, E. Kawasaky, H. R. Bourne & F. McCormick, unpublished work). Although the biochemical effect of these mutations on α_i2 function and the mitogenic pathway involved remain to be identified, it appears most likely that the *gsp* oncogene will turn out to be only the first example of a new class of oncogenes resulting from mutations in the genes that encode signalling G-proteins α-chains.

References

1. Stryer, L. & Bourne, H. R. (1986) *Annu. Rev. Cell Biol.* **2**, 391–419
2. Neer, E. J. & Clapham, D. E. (1988) *Nature (London)* **333**, 129–134
3. Casey, P. J. & Gilman, A. G. (1988) *J. Biol. Chem.* **263**, 2577–2580
4. Berridge, M. J. & Irvine, R. F. (1989) *Nature (London)* **341**, 197–205
5. Dumont, J. E., Jauniaux, J. C. & Roger, P. P. (1989) *Trends Biochem. Sci.* **14**, 67–71
6. Billestrup, N., Swanson, L. W. & Vale, W. (1986) *Proc. Natl. Acad. Sci. U.S.A.* **83**, 6854–6857
7. Vallar, L., Spada, A. & Giannattasio, G. (1987) *Nature (London)* **330**, 566–568
8. Landis, C. A., Masters, S. B., Spada, A., Pace, A. M., Bourne, H. R. & Vallar, L. (1989) *Nature (London)* **340**, 692–696
9. Masters, S. B., Stroud, R. M. & Bourne, H. R. (1986) *Protein Eng.* **1**, 47–54
10. Barbacid, M. A. (1987) *Annu. Rev. Biochem.* **56**, 779–827
11. Holbrook, S. R. & Kim, S. H. (1989) *Proc. Natl. Acad. Sci. U.S.A.* **86**, 1751–1755
12. Masters, S. B., Miller, R. T., Chi, M., Chang, F., Beiderman, B., Lopez, N. G. & Bourne, H. R. (1989) *J. Biol. Chem.* **264**, 15467–15474
13. Graziano, M. & Gilman, A. G. (1989) *J. Biol. Chem.* **264**, 15475–15482
14. McCormick, F. (1989) *Cell (Cambridge, Mass.)* **56**, 5–8
15. Graziano, M. P., Freissmuth, M. & Gilman, A. G. (1989) *J. Biol. Chem.* **264**, 409–418

Subject Index

First and last page numbers of papers to which entries refer are given